T0332231

CONSCIOUSNESS AND ROBOT SENTIENCE

Second Edition

Series on Machine Consciousness
ISSN: 2010-3158

Series Editor: Antonio Chella *(University of Palermo, Italy)*

Published

Vol. 4   *Consciousness and Robot Sentience (Second Edition)*
         by Pentti O Haikonen

Vol. 3   *The Revolutions of Scientific Structure*
         by Colin G Hales

Vol. 2   *Consciousness and Robot Sentience*
         by Pentti O Haikonen

Vol. 1   *Aristotle's Laptop: The Discovery of our Informational Mind*
         by Igor Aleksander and Helen Morton

Series on Machine Consciousness – Vol. 4

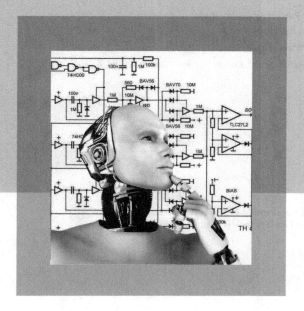

CONSCIOUSNESS AND ROBOT SENTIENCE

Second Edition

Pentti O Haikonen

University of Illinois at Springfield, USA

W **World Scientific**

W JERSEY • LONDON • SINGAPORE • BEIJING • SHANGHAI • HONG KONG • TAIPEI • CHENNAI • TOKYO

Published by

World Scientific Publishing Co. Pte. Ltd.

5 Toh Tuck Link, Singapore 596224

USA office: 27 Warren Street, Suite 401-402, Hackensack, NJ 07601

UK office: 57 Shelton Street, Covent Garden, London WC2H 9HE

British Library Cataloguing-in-Publication Data
A catalogue record for this book is available from the British Library.

Series on Machine Consciousness — Vol. 4
CONSCIOUSNESS AND ROBOT SENTIENCE
Second Edition

Copyright © 2019 by World Scientific Publishing Co. Pte. Ltd.

ISBN 978-981-120-504-0

For any available supplementary material, please visit
https://www.worldscientific.com/worldscibooks/10.1142/11404#t=suppl

Desk Editor: Tay Yu Shan

Printed in Singapore

Dedication

This book is humbly dedicated to you, my respected reader and to the international community of machine consciousness researchers.

Preface

Contemporary AI has a hidden problem. This problem prevents AI from becoming what it is supposed to be in the future as foreseen in science fiction. There will be no robots that understand what they are doing, unless this problem is solved. And it is solved here.

THIS BOOK is the fully revised and updated second edition of the author's book "Consciousness and Robot Sentience". This edition has lots of new material, and will provide new insights that go beyond the first edition. The organization of this book is revised and streamlined for better clarity and continuity of the lines of arguments.

The viewpoint of AI has been added to this edition. It is shown that AI has a hidden problem, and this problem prevents AI from becoming a true agent that understands what it is doing.

A self-evident solution to the hidden problem of AI is found and given. It turns out that the solution to AI's hidden problem is surprisingly connected with the concepts of qualia, the mind-body problem and consciousness. These are the hard problems of consciousness that so far have not been solved properly. Unfortunately, the solution to the hidden problem of AI cannot be satisfactorily implemented, unless the phenomena of qualia and consciousness are understood.

In this book the hard problems of consciousness are addressed, and an explanation of consciousness is presented, one that rejects material and immaterial substances, dualism, panpsychism, emergence and metaphysics generally. What remains is obvious, and also allows the artificial creation of conscious machines.

However, the given explanation of consciousness excludes consciousness in digital computers, but allows the artificial creation of consciousness in one natural-like way, by associative non-computational neural networks.

The proof of a theory is in its empirical verification. In this case, the final proof could be in the form of a sentient robot. This book describes a step towards this in the form of the author's small experimental robot XCR-1. This non-digital, neural hardware robot has evolved through the years, and has now new cognitive abilities, which are described.

Finally, I want to thank the publisher for the possibility to write this book and the good people at World Scientific for their kind, expert efforts.

I also want to thank you, my respected reader for your interest in my modest work.

I thank also my wonderful wife Sinikka for her support and patience. Special thanks go to my media artist son Pete Haikonen for the production of the related XCR-1 robot demo videos[1] with exciting original music.

March 31, 2019

Pentti O A Haikonen

---

[1] XCR-1 robot demo videos here: https://www.youtube.com/user/PenHaiko

Contents

Chapter 1

Artificial Intelligence

1.1. AI, Computation and Cognition

Is it possible to make a computer to think? Are computers thinking already? These are old questions. Already in the early days of computers, in the 1950's, some researchers thought that they had the answers ready. When humans compute, they think. What does the computer do, when it executes exactly the same computations? Humans can also reason non-numerically. What does a computer do, when it is programmed to do the same reasoning exactly in the same way? Wouldn't it be fair to say then that the computer thinks? The early researchers thought so, and this hypothesis gave rise to the discipline of Artificial Intelligence (AI).

The fundamental assumption behind AI is that both thinking and cognition are computational and symbolic, and therefore can be produced via the execution of algorithms. This view leads to the conclusion that human-like general intelligence can be produced by suitable computer programs, and eventually, the computer should be able to think and reason as well or even better and faster than a human. It would all depend on the extent and ingenuity of the programs.

However, all is not well, and the foundations of AI are not so solid as they are made to appear. AI has a fundamental, embarrassing problem that everybody knows, but nobody wants to talk about. Yet, this problem prevents AI from becoming what it is supposed to be. Also, it turns out that the studying of this problem will reveal an unavoidable connection between intelligence and consciousness; there cannot be any true intelligence without consciousness, as will be pointed out later on.

The fundamental problem of AI was not initially recognized by the AI pioneers, and when it eventually was, it was denied, belittled and played down. Still, after sixty plus years after the first AI programs, AI is haunted by this problem, and the situation is not getting better. In fact, it is getting worse and straightforward dangerous, as more complicated AI programs with autonomous executive powers are being fielded.

How did the fundamental problem of AI arise? Artificial Intelligence was born in mid-1950's when Herbert Simon and Al Newell produced their first AI program, the "Logic Theorist". This program was different from all previous computer programs, as it did not do numerical computations. Instead of these, it executed logical reasoning. Now it appeared that computers could be more than mere programmable numeric calculators. Herbert Simon claimed that he and Al Newell had invented a thinking machine, and in doing so they had also solved the mind-body problem that had puzzled the philosophers of the mind for eons. Later on Simon and Newell presented their "Physical Symbol System Hypothesis" (PSSH) that became to be the cornerstone of Artificial Intelligence. According to this hypothesis a rule-based symbol manipulating computer has everything that is necessary for general intelligence. Thus, thinking and intelligence are nothing more than rule-based symbol manipulation, and a suitably programmed computer would eventually be able to execute every mental operation that is executed by the human mind and brain [Newell and Simon 1976].

Simon and Newell were not able to verify experimentally their Physical Symbol System Hypothesis, apparently due to a practical reason, namely the limited processing power and memory capacity of the computers of that era. Instead of a direct proof they proposed that the hypothesis was actually verified by indirect evidence, the fact that there were no known other means and mechanisms for thinking and cognition. If thinking were not rule-based manipulation of symbols, then what else could it be? Nothing else. This is the only way and "there is no other game in town" [Fodor 1975]. This conclusion was accepted at its face value by the forthcoming AI researchers, even though it was based on a logical fallacy; *argumentum ad ignorantiam*, the ignorance of evidence to the contrary. Unfortunately, the ignorance of any contradicting evidence is not a proof of the non-existence of such evidence, it is a proof of something else. Simon, Newell and Fodor could not think of any other explanation for the processes of thinking, but this ignorance does not constitute any logical proof for their hypothesis.

The digital computer is a physical symbol system, where binary words, strings of zeros and ones, are used as the symbols. Computers are known to work very well, and they are able to perform a wide variety of information processing tasks, also those that apparently call for some kind of intelligence. For instance, computers can successfully play games and drive self-driving cars. No doubt, even more astonishing applications will be seen. So what, if any, is the problem with physical symbol systems?

There is a serious problem, and it will be shown here that the Physical Symbol System Hypothesis is not valid. The brain is not a digital computer, and it is not a physical symbol system even though it is able to think and reason in symbolic ways. There is a fundamental difference between the ways in which information is processed in the brain and the computer, and this difference prevents the creation of computer-based true intelligence. In the following this difference is explained, and it is also explained how the problem can be remedied and how true thinking machines can be designed.

1.2. The Difference between the Brain and the Computer

Complicated calculations involve the serial execution of different mathematical operations and the storage and reuse of intermediate results. First computers were designed for the automatic execution of strings of numeric calculations. They were calculating machines with memories for intermediate results and the type and order of the operations to be executed; the program. In addition to the calculating unit and memory, a special control unit was needed to control the overall operation. Contemporary computers are vastly refined, but the basic principle of the combination of program, calculator, control and memory is still the same. Without programs computers do nothing.

A computer memory consists of addressable memory locations for each piece of data, which is in the form of binary words. The running of a computer program involves data retrieval and storage with the help of memory addresses. This can be demonstrated by a trivial example of a bank account balance computation command:

balance = balance + deposit

This command states that the numeric values from the memory location addresses "balance" and "deposit" must be added together, and the sum must be stored to the memory location address "balance". Now a computer novice may claim that *balance* and *deposit* are not memory addresses, they are names of variables. This is how it looks, and it might also look like the computer would actually understand, what the computation is about. However, *balance* and *deposit* are only labels for the actual memory location addresses, which are binary numbers. The numeric values stored at these memory locations are the "variables" that may change. The computer reserves a memory location with an address for each variable whenever the program is to be run.

The labels *balance* and *deposit* do not carry any external meaning to the computer, but are helpful for the programmer and anyone trying to figure out what the program does. The stored numeric values of variables do not have any external meaning to the computer, either. The running of a computer program involves the handling of memory location addresses, not external meanings.

The brain has no addressable memory locations, and consequently, no memory address handling and management is required. Instead, the brain operates with phenomenal meanings, produced by the senses and retrieved from memory. Memories are evoked by "mental images", and in this sense the information itself is also the "memory address". Information processing and memory function are seamlessly combined in the brain. The flow of mental action is not controlled by any programs, instead it is driven by internal and external conditions and situations.

It should be obvious that the operational principles of the computer and the brain are completely different. The human mind operates with meanings, but where is the meaning in the computations of the computer? This question relates to the so called symbol grounding problem. Meanings cannot be used in computations, if this problem is not solved.

1.3. Meaning in Symbol Systems

Let's suppose that you are in captivity, sitting inside a windowless room. You have no idea how you got there, and you do not know what is outside. There is a monitor and a calculator in front of you. In order to get food you have to use the calculator to do given computations with numbers that appear on the monitor screen and type the results to the system. Eventually you learn to do this quickly, even though you have no idea what the numbers are about. Then, suddenly the door is kicked open, police officers rush in and you are taken to court and charged with homicide. It turns out that your computations have actually controlled a self-driving car. There has been an accident, and a passenger has died. You try to explain that you are not guilty, because you have not been able to understand what you have been doing, as you have had no way of knowing what the numbers and calculations mean. The prosecutor is not impressed maintaining that all this is irrelevant. Your operations had been successful for a good while, and from the outside it has appeared that you have understood what you have been doing. What else could be required?

Without meanings there can be no understanding. Without understanding there can be no true intelligence. Rule-based computations will not reveal what the numbers mean and are about. Syntactic manipulation of symbols will not lead to semantics, and the meanings of the symbols will not be revealed in this way. American philosopher John Searle tried to point this out with his famous "Chinese Room" thought experiment, where a non-Chinese person inside a room answers written Chinese language questions in written Chinese with the help of rules and look-up tables [Searle 1980]. From the outside it appears that the room or somebody inside the room was able to understand Chinese language symbols, but it is known from the set-up that this is not the case.

Searle explained that computers are kinds of Chinese rooms, operating blindly with rules and symbols without meanings, and therefore are inherently unable to understand anything. This argument did not go well with Strong AI enthusiasts, who maintained in the good tradition of the Physical Symbol System Hypothesis that a suitably programmed computer with proper inputs and outputs will have a mind in the same sense as the humans have. Searle did not accept this, and argued that understanding will not arise in the computer no matter what kinds of rules are programmed, because the external meanings of the symbols are neither accessed nor utilized. In the Chinese room information is processed by blind rules, not by meanings, and the same goes for computers, too.

Searle's argument is related to the so called *symbol grounding problem*: How the meanings of symbols can be defined and incorporated in symbol systems [Harnad 1990]. A symbol in itself is only a distinct pattern with no inherent meaning. Words and traffic signs are everyday examples of symbols. If you have not learned what they mean, they are meaningless to you.

The symbol grounding problem is also apparent in the case of dictionaries. A good dictionary defines the meaning of every word with the help of other words. Thus, it would appear that the symbol grounding problem is solved there. This is not the case. When one looks for the meanings of the words that explain the word to be explained, one eventually ends up in a circle, where the explaining words are explained by the word to be explained. For example, Webster's New World Dictionary from the fifties defines "red" as the color of blood. Ok, but what is "color" then? Webster knows: colors seen by the eye are orange, yellow, red... and you are no wiser.

Mathematics is an other symbol system affected by the symbol grounding problem. Let's consider a simple example.

Let $A = 5B$. What is the meaning of B? This can be solved by the rules of algebra and we get $B = 0.2A$. But did we get the real meaning of B? No. The meanings of A and B remain unsolved, and cannot be revealed simply because they are not there.

It should be evident that mathematical rule-based operations will only tell and reveal something about the relationships between the symbols used in the computations, but nothing about their actual intended meanings. There is no such mathematical operation that could reveal any external meanings, and these meanings, if any, remain only in the mind of the person doing the calculations.

In physics, physical units such as the meter, second and kilogram are carried with the equations. At first sight it might appear that in this way meanings were attached to the calculations. However, this is not the case, and the symbol grounding problem is not solved. The unit markings are just letters and may be used in the algebraic computations in the same way as the other symbols in the equations, in the way of dimensional analysis. No meaning is carried into, or captured by the process of computation or the computing system itself, and the understanding of the external meanings remain for the human supervising the calculations. The very universality and power of mathematics arises from the fact that meanings are omitted. It does not matter what is counted; beans, bananas or money. But, from this universality it also follows that the numbers and calculations alone will not reveal what are being counted.

The lesson here is that in a symbol system the meanings of symbols cannot be ultimately defined by other symbols in the system, nor can they be revealed by any computation. At least some of the meanings must be tied to and imported from the outside world. Therefore, a system that operates with meanings must be able to acquire external information. Humans have senses for that purpose, and nowadays also computers can be fitted with cameras, microphones and any other sensor that an application requires. Thus, it should be technically possible to solve the symbol grounding problem.

However, there is an unfortunate complication. Meanings cannot be imported in the form of symbols, as the imported symbols would only increase the number of symbols to be interpreted. Therefore, the meanings must be imported in a form that requires no interpretation; in *self-explanatory forms of sensory information*. Symbols are not these.

This requirement leads to another catch: A conventional symbol system is able to handle symbols only, as there is no provision for any other form of expression. Non-symbols cannot be accommodated. A digital computer is able to accept only binary words as its input. Consequently, any analog input, such as audio or vision information must first go through analog-digital conversion. This conversion outputs binary numbers, which are symbols. As such they require interpretation, and no symbol grounding has been achieved. This is a serious problem that leads to the fundamental problem of AI.

1.4. The Fundamental Problem of AI

The fundamental problem of Artificial Intelligence arises from the fact that computers do not operate with external meanings, they operate only with blind rules and naked data. Without meanings there cannot be any understanding, and without understanding there cannot be any true intelligence. Therefore contemporary AI is not true intelligence.

The computer is a symbol system, operating with programmed rules and binary word symbols. The possible external meanings of the symbols are not available without interpretation, and in practice this interpretation is done by the human using the computer.

The external meanings must be imported into a symbol system, and this calls for external information acquisition and suitable sensors. However, there is a problem. The imported meaning cannot be in the form of a symbol, because this action would only increase the number of symbols to be interpreted. Unfortunately a digital computer cannot accept any information in other forms than symbols, and therefore the symbol grounding problem cannot be solved, as symbols cannot be ultimately interpreted by other symbols only. This means that the digital computer cannot operate with meanings in the true sense, and consequently it will not be able to understand what it does. Robots with symbolic AI do not understand what they are doing. They may appear to converse fluently with humans, but in reality they do not know what they are talking about. They do not even know that they exist.

True Artificial Intelligence calls for different kind of information processing machinery. This machinery would be able to perceive itself and its environment, and for the grounding of meaning of symbols it would use sensory information in such self-explanatory forms that do not require interpretation.

These conclusions lead to other questions: Meanings must be imported in self-explanatory forms of information, but what exactly would these never heard of forms be? Has anyone ever seen this kind of information, for that matter? The answer is obvious, and can be found by inspecting the process of sensory information acquisition. It will then also turn out that the issue of self-explanatory information is related to the problem of consciousness.

Chapter 2

Sensory Information and Meaning

2.1. Sensing and Meaning

The human mind acquires all its experience about the environment and the body by the help of various sensory channels, such as visual, auditory, touch, etc. Sensory systems use *receptors* or their technical counterparts, *sensors*, that are sensitive to the specific external stimuli of their kind. In the visual modality photons are sensed, in the auditory modality vibrations of air pressure are sensed, in the touch modality variations of contact pressure are sensed.

Receptors and sensors are *transducers* that convert the external physical stimuli into the specific common physical form that is used inside the system. In the brain the response is in the form of neural signal patterns, in the computer the response is the form of electric binary words. In other electronic information processing machines the response may be in some other electric signal pattern form. Examples of artificial transducers are microphones, image sensors, pressure sensors, etc. Figure 2.1 depicts the general sensing process that can be either analog or digital.

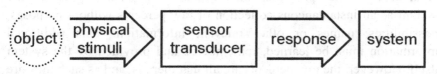

Fig. 2.1. The sensing process. The sensor receives a physical stimuli pattern from the sensed object. This stimuli pattern evokes a response pattern in the sensor/transducer. This response pattern is in the physical form that is accepted by the system. In analog systems the transducer maps stimuli patterns into corresponding response patterns.

In Fig. 2.1 the sensor receives physical stimuli, such as reflected light, from the sensed object, or the stimuli itself may be the sensed entity, such as sound. The sensor reacts to these stimuli and generates response patterns in such a physical form that is accepted by the system. The response pattern is the only thing that the system "sees". These response patterns may also be dynamic, rather than static "images".

The logic of a properly working analog sensing process is simple: If a certain stimuli pattern is present, then a certain response pattern is present. This works also the other way round: If a certain response pattern is present, then a certain stimuli pattern is present. In this way the transducer maps stimuli patterns into corresponding response patterns. Thus, the response patterns have a meaning; their presence indicates the presence of their corresponding stimuli; they stand for the stimuli and the cause of the stimuli. In this way the external world is imported into the system in an analog way, in the terms of sensory response patterns.

In digital sensing systems the response is in the form of binary numbers. No mapping takes place, and the external world is not imported into the system. Digital sensing does not readily convey meaning.

Analog sensing process produces signal patterns that stand for the sensed entities. The information that these patterns convey is simple and self-explanatory. A seen round thing is a seen round thing. A felt round thing is a felt round thing. A heard sound is a heard sound. The experience of red is red. These sensory response patterns are called here raw percepts.

Sensing is useless, if the sensory responses cannot be utilized. Without any other processes the sensed world would appear to the system as an instantaneous collection of raw percepts without purposes, possibilities, names or other relevant information. This additional information must be learned, remembered and provided by the system itself. However, the raw percepts are all that there is, and as such they are only available, when the corresponding stimuli are present. This would be good enough for simple instantaneous stimulus-response reactions, but not sufficient for deliberated responses or symbolic thinking.

Therefore additional mechanisms and processes are required. The system has to be able to disconnect raw percepts from the presence of their evoking stimuli by learning, memorizing and recalling them when needed, and be able to use them also as symbols for cognition. These symbols would have associated meanings, fundamentally grounded to the external world via the self-explanatory sensory response patterns. In principle, this can be achieved in a way outlined in the following.

2.2. Learning of Meaning

As themselves, raw percepts have only one meaning, namely the indication of the presence of the sensed entity. Even the simplest animal needs to know more. In order to survive it has to know, what a raw percept is about; is it dangerous, is it edible, is it good for mating, is it useful in some other way, can it be ignored. These meanings must be learned and remembered.

For a coherent world model, raw percepts must be connected. Percepts of dangerous objects must be associated with pain, percepts of edibles must be associated with eating, and so on. The acquisition of language calls for the association of names with perceived entities.

Associative learning is the act of connecting percepts with each other in such a way that one of the connected percepts is able to evoke the percept of the other, also in the absence of any related external stimuli. In the brain this operation is performed by the neurons.

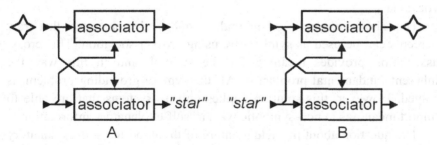

Fig. 2.2. Cross-association of raw percepts attaches meanings to these. The raw percept of the sound pattern "star" and the raw percept of the star pattern are associated with each other by their simultaneous presence. Afterwards the star pattern is able to evoke the raw sound pattern "star" (A), and the sound pattern "star" is able to evoke the star pattern (B). Cross-association also disconnects raw percepts from the sensor stimuli, so that the percepts can be available for the system regardless of the external situation.

Figure 2.2 gives an example of the connection of raw percepts via cross-association. In this process simultaneously appearing raw percepts are associated with each other, so that after the learning of this association one of the raw percepts is able to evoke the other. Cross-association gives associated meanings to percepts, and also provides the transition from sub-symbolic operation to symbolic operation. Cross-association also disconnects raw percepts from the sensor stimuli, so that the percepts can be available for the system regardless of the external situation. In this way cross-association provides the function of memory.

A natural language uses cross-association for the grounding of meaning of the used words. Initially the heard words are only sound patterns without any meaning. Anyone can verify this by listening to an unknown foreign language. The grounding of meaning of the words begins by the cross-association of the heard sound pattern with a percept of an entity. Here the raw percept of a sound pattern in itself is not a symbol, it is just a self-explanatory percept of sound with no further meanings. However, after the cross-association it has become to be a symbol, a word, that stands for the associated entity. The cross-association works both ways. A name can evoke a memory of the associated entity, and the entity can evoke its name. In this way a basic vocabulary of grounded words is created. After this point, the meaning of further words may be defined indirectly by using the words of the vocabulary.

The above example of natural language demonstrates that raw percepts can be used as symbols by using cross-association. This cross-association provides meaning for the symbol, and in this way the inherent fundamental problem of AI, the symbol grounding problem, is solved. However, this solution applies only to systems that are able to import meanings in non-symbolic way, as self-explanatory information.

The question about the actual nature of the imported self-explanatory information remains. This will be discussed and its relation to consciousness will be revealed in the following.

Chapter 3

Self-Explanatory Information and Qualia

3.1. The Appearance of Self-Explanatory Information

In order to be self-explanatory and require no interpretation, sensory information would have to have some kind of internal appearance of external forms, gestalts, patterns, properties and qualities. In this way it would be able to import an impression and appearance of the external world and body into the system.

In reality, this is indeed the exact experience that our senses produce for us. The neural activity produced by the senses has internal appearances of what we see, hear, smell, taste, feel. This is self-explanatory sensory information, because it appears directly as the observed entities, needing no interpretation. In the philosophy of mind the experienced qualities produced by sensors are known as qualia (plural, singular quale). Qualia are self-explanatory information.

The word "qualia" refers to the phenomenal feel and quality of sensory percepts. Typical qualia are, for example, the perceived colors, the timbre of sounds, the taste of edibles, the wetness of water, the softness of a tissue, the coldness of ice, the feel of pain and pleasure. Qualia are the subjective qualities of percepts. This is different from digital data, which by its very nature does not have any appearance.

Despite their appearance, most qualia are not directly real world properties, even though they have a causal connection to these via the sensing process. In reality, the appearances of qualia are based on the reactions of the sensors and the system. For instance, atoms and molecules do not have any intrinsic color, odor or taste to be readily perceived. Instead, the impressions of these are produced by sensor reactions to the actual sensed stimuli. Different photon energies cause

13

impressions of different colors. Molecule structure and chemical activity cause experiences of taste and odor. This is an obvious consequence of the sensing process, but there is also other evidence. For example, visual after-image effects that can be seen also in total darkness, show that qualia are caused by the sensor itself, as no external stimuli are present.

Qualia have a special property: They can be named, but not described. For example, describe blue. "Blue is the color of the sky". Yes, sky is blue, but this is a description of the sky, not the experience of blue. Likewise, describe sweet taste. "Sweet taste is the taste of sugar". Yes, sugar has sweet taste, but again, this is a description of sugar, not the sweet taste. Qualia are primary experiences, and there is nothing more primary that could be used for their description. An attempted description of a quale will not amount to the understanding of that quale for a person who has not experienced the quale in the first place.

Qualia are subjective, because their phenomenal essence cannot be communicated by descriptions. We know the appearance and feel of our own qualia, but we do not know, if others have qualia, and what kind of qualities these would have. We do not know, how other people perceive the world. However, the biological similarity of people leads us to assume that the qualia of other people are similar to ours, but this is not a proof. The situation is illuminated by the condition of color blindness. Obviously colorblind people experience colors in a different way. For instance, they may perceive the colors red and green as the same color, but which color would that actually be? Would it be the red or the green of the people with the "normal" vision? Or would it be a some kind of brown? But, what would be the color of the "brown" of colorblind people? And, would your "red" be similar to mine, for that matter? How can we know, and does it really matter?

Qualia are internal appearances in the brain, and in order to see other people's qualia we should somehow connect their brains directly to ours, so that their qualia-generating brain activity would generate corresponding qualia in our brain. However, even then the problem would not be solved, as the generated qualia would be ours. There is no guarantee that these qualia would be similar to the other person's qualia.

Qualia remain non-describable, but we can use qualia to describe things, because usually qualia are common experiences between us.

However, qualia can be reproduced by the reproduction of the corresponding sensory stimuli, like in television.

It is also known that the artificial excitation of sensory nerves produces qualia. This fact is already utilized by the cochlear implant technology. A cochlear implant is an electronic device that is surgically implanted into the inner ear of a deaf patient with a defunct cochlea. A normally functioning cochlea resolves the incoming sound into its constituting frequencies and transmits corresponding neural signals to the brain. A cochlear implant does the same thing by receiving sound via a microphone and stimulating the auditory nerves electrically according to the spectral content of the sound. This results in the perception of sound-like qualia and a patient with a cochlear implant will understand speech and even hear music. Here the auditory qualia are clearly generated by the artificial electric excitation of the auditory nerves.

Thus, sensory qualia arise from the sensory organs' responses to stimuli and are carried by the excitation of sensory nerves. The question remains, how these signals and signal patterns would appear as different qualia.

3.2. Different Qualities of Qualia

All sensory modalities produce percepts with qualia that are typical to the producing modality. Yet, in each case the experience is conveyed by neural signals that seem to be similar to each other. It is obvious that qualia must be different, how else could they depict the different qualities of the world. The question is: How can this differentiation be achieved by using similar neural signals?

In color television the colors are reconstructed by using three primary colors; red, green and blue (RGB). In analog studio the RGB color information is carried by three separate lines, one line for each color. The signals in these lines are superficially similar to each other and an external observer may not necessarily be able to identify each color by merely inspecting these signals. In this case the differentiation is realized by the hardware, the fixed wiring paths of these signals.

In the brain a similar scheme seems to be used. The quality of the evoked qualia is tied to the identity of the nerves. Auditory qualia arise from the stimulation of the nerves that originate from the cochlea and terminate at the auditory cortex. Likewise, the stimulation of nerves that go to the visual cortex cause visual qualia, etc. This specificity of nerves can be demonstrated in various ways. For instance, a local push on the skin causes a tactile sensation, while a similar push on the eye leads to the stimulation of visual nerves and causes visual sensations.

Thus, the evoked qualia seem to be determined by the fixed neural wiring. A certain sensory nerve would be associated with a certain quale. Obviously this is very much so, but this does not yet explain, what it is exactly that makes the quale appear different. A certain nerve carries by the variable pulse repetition rate the qualia of "reds", while the other one carries the qualia of "greens" for a given pixel in the retina. These qualia have a specific appearance, which is different from each other, but why? In technical terms fixed neural wiring would be sufficient for the discrimination of colors (and other sensory information), but the fixed wiring alone will not explain the appearance of the subjective experience.

This problem was recognized by Johannes Müller (1801–1858), and his theory known as *the Law of Qualities* tried to explain that [Gregory 1998]. According to that theory the quality of the evoked qualia is not generated at the origination points of the sensory nerves (the sensor), instead it is generated at the end points of these nerves, namely at the target area in the cortex. Each sensory modality has its target area in the cortex and this, according to Müller, is where the experience of qualia arises as a response to the stimuli carried by the incoming sensory nerves. Thus, the subjective experience of the quality of a quale would not be carried by the nerve signal, instead it would be created in the specific cortical areas for each sensory modality. Obviously these specific cortical areas would have to adapt to the type of information that they are receiving, like the serial auditory information and the spatially parallel visual information. The created qualia would have to reflect the format of the incoming information. Unfortunately Müller's theory does not help much, because the mere statement that something happens somewhere is not an explanation.

Another experiment illustrates the phenomenon of qualia. Take an audio signal generator that is able to produce sinewave frequencies between 10 Hz and 20 kHz. Listen to the signal with good earphones, and start from 10 Hz. At that low frequency each cycle of the sound can be heard separately, and in fact, the sound is perceived more like a monotonous rhythm. Increase then the frequency. You will notice that beyond 100 Hz the separate cycles fuse together, and the experience is that of a whistle. It is easy to understand how very low audio frequencies are perceived as kinds of rhythms, because the pulse repetition rate of neurons can easily follow the audio cycles. However, it is not so clear why higher frequencies should be heard as higher and higher pitched whistles, when the maximum pulse repetition rate of neurons has been exceeded.

3.3. Amodal Qualia

It was stated before that most qualia are not some properties of the real world. This statement needs to be clarified here. There are certain detectable features that are the same in several sensory modalities, such as the seen and heard direction of a sound source, felt and seen shapes and, especially, rhythm and interval duration. For example, feeling the teeth with the tip of the tongue results in shape percepts that are image-like. These properties are called *amodal features*. It is argued that these, when perceived, appear as *amodal qualia*. (This is not to be confused with the so-called amodal perception, e.g. the virtual perception of a partly hidden object as a whole.)

Amodal features are invariants; they are the same in the observed world and in the resulting neural activity. For instance, the rhythm of a piece of music is a feature of the actual sensed phenomenon, and it should also be a feature of the resulting neural activity. Also changes in percepts, like spatial and temporal intensity changes etc. can be depicted in amodal ways.

It was also stated before that, in general, we cannot know what kind of qualities other people's qualia would have. However, amodal qualia may be an exception to this rule. For instance, rhythm is the same for everybody, and can be considered as a *shared quale*. In the case of cognitive robots this leads to an interesting result: We could know how an amodal quale would appear in a robot, if the robot had internal appearances in the first place. Nagel [1974] argued that we cannot possibly know how it feels to be a bat. However, it would seem that amodal qualia could allow us to know to a certain extent how it feels to be a bat. Nevertheless, without additional information it is not possible to know that the bat would actually feel anything.

All features are not fundamentally amodal. Examples of features that are seemingly not amodal include colors, taste, smell, etc. However, these features may appear in the context of amodal features.

Amodal features offer also a natural way of sensorimotor integration. An example of sensorimotor integration is dance, where the rhythm of the music is also the rhythm of the motion. It is obvious that the perceived amodal feature, the rhythm, can be directly used as motor control commands, and it can also be shared with other dancers. Also the production of speech would seem to utilize amodal features.

3.4. The Fundamental Problem of Qualia

Previously it was stated that qualia arise from the sensory organs' reactions to stimuli. These reactions generate neural signals and signal patterns that are transmitted to the brain. In principle these signals can be monitored by existing laboratory instruments, but in practice the skull would have to be opened for the access to individual neurons. There are other indirect non-invasive means of recording brain processes with various degrees of accuracy. These methods allow the detection of qualia-related electric and electromagnetic fields and potentials on the scalp and spatial metabolism changes inside the brain. The detected brain activity can be found to be associated with certain mental events by the subject's reports and some indirect ways. However, so far no qualia or any mental state has been detected directly as such. Yet, it is obvious that all information in the brain is carried by the brain's neural processes. These processes are physical. However, no qualia or any internal phenomenal appearance in the brain have been detected by physical instruments. Therefore, would it be possible that qualia are not physical at all?

Jackson [1982] argued that qualia are not physical. According to Jackson, all the information related to a percept can be described without qualia, and therefore, the inclusion of the "feel" of qualia does not add any physical information. In his famous "Mary argument" Jackson presented a scientist, Mary, who knew everything about seeing colors; all the reactions in the eyes and in the brain, etc. However, Mary had been living in a black and white room for all her life, and thus had never actually seen any colors. However, she knew what it would be like to see a color. Therefore the actual act of seeing of a color, the perceiving a color quale, would not increase her knowledge about seeing colors. Thus, the addition of qualia would not increase the amount of the physical information that Mary already had about colors, therefore qualia would neither be physical nor necessary, obviously.

It is true that the information related to qualia can also be represented by indirect ways. This is demonstrated by Fig. 3.1, where color information is presented by symbolic labels.

Fig. 3.1. An image with indirect symbolic description of color information. In order to understand the color information one has to know what the labels mean. Colors would be self-explanatory, if the picture were printed in color.

The labels in Fig. 3.1 tell what kinds of color qualia you would experience, if the picture were printed in full color and you were able to see colors. More accurate color description could use numeric information that would indicate the value of the hue and intensity of each pixel. In this way, the labels are symbols for the corresponding qualia. These symbols, labels or numbers, in themselves do not contain any self-explanatory information. Therefore they must be interpreted by learned or otherwise available knowledge. This knowledge would have to be grounded to actual sensory color experience.

Jackson's "Mary argument" is wrong. For the argument's sake, let us suppose that you are told, how it feels to put your finger in the candle flame. The flame will burn your finger badly, you will withdraw your hand immediately, your heartbeat will increase etc. In short, you would know exactly, what would happen. The actual execution of this act would be accompanied by qualia, but technically, you would not learn anything new that you did not already know. So, what would be the contribution of qualia, if they did not increase your knowledge? It is the experience. Information without qualia is just a description; you have been told how it feels, but you have not actually felt it.

If the picture of Fig. 3.1 were printed in color, color qualia could be directly and subjectively experienced requiring no further descriptions, symbols or commonly agreed comparison values. Red is red, sweet is sweet and pain is pain; we do not have to learn their names in order to experience them as they appear. We do not have to learn them via associations with some other entities, either. Qualia are not symbols, because they do not require any reference to some common knowledge in order to be understood. Instead, qualia are a direct, self-explanatory way of experiencing information. Instead of description, qualia are physical experience; you, your body and senses experience the situation and react to it, you actually feel and live through it. Without qualia the grounding of meaning remains missing.

A difficult problem remains: How can qualia appear in an apparently immaterial way as qualities of the world, instead of appearing as the neural activity that they really are? This problem is related to the *hard problem of consciousness*. Qualia solve the symbol grounding problem, but in order to solve the problem of qualia, the problem of consciousness must be solved, too. This is done in the following two chapters.

Chapter 4

Hypotheses about Consciousness

4.1. The Mystery of Consciousness

Ancient philosophers noted already thousands of years ago that there is something special in being a conscious human. What is it that makes us able to think and reflect and be aware of the world and our own existence? What is the impression that we are a conscious entity, a soul inside our body?

It is easy to get the impression that mind and body are different. When we work, we see our body moving, and we may get tired. This is definitely something material. However, when we think, we do not perceive any physical processes taking place at all, and consequently our mind appears to be immaterial. The Greek philosopher Plato argued that this separation is real and two different worlds exist. According to Plato the body is from the material world, and the immaterial soul is from the world of ideas. Plato's explanation was widely accepted, and his ideas were further developed by many later philosophers, like the French René Descartes.

4.2. Cartesian Dualism

René Descartes suggested in his book *The Discourse on the Method* (1637) along the lines of Plato that mind and body are of different substances. The body is a kind of a material machine that follows the laws of physics. The mind, on the other hand, is an immaterial entity that does not have to follow the laws of physics, but is nevertheless connected to the brain. According to Descartes, the immaterial mind and the material body interact; the immaterial mind controls the material body, but the body can also influence the mind. This view is known as

Cartesian dualism. Cartesian dualism is also *substance dualism* as it maintains that two different substances, material and immaterial ones exist.

Is the dualist interpretation correct, and do we really have an immaterial mind? This is a crucial issue for machine consciousness. Conscious machines should also have a mind with the presence of apparently immaterial subjective experience, more or less similar to ours. The artificial mind should thus contain mental content, which would appear to the machine as immaterial, without the perception of any underlying material processes. Machines are material; therefore, if real minds were immaterial, then the creation of immaterial machine minds by existing material means would be impossible. The dualist view of mind is illustrated in Fig. 4.1.

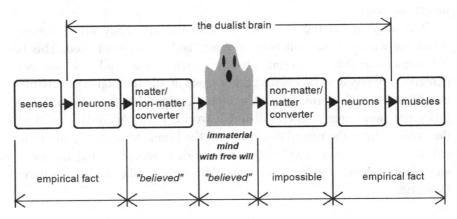

Fig. 4.1. The dualist view of mind. Senses provide information about the material world in the form of material neural signals. These affect the immaterial mind, which, in turn controls the material neurons that operate the muscles. In this process a non-matter/matter converter is needed. This converter would violate physics as it would have to produce matter and/or energy out of nothing.

The dualistic approach may at first sight seem to succeed in explaining the subjective experience of the immaterial mind with the proposed existence of two different substances; the material substance of the physical world and the immaterial substance of the mind.

Unfortunately, while doing so it leads to the *mind-body interaction problem*; how could an immaterial mind control a material body?

It is known that the brain receives information about the material world via sensors; eyes, ears, and other senses that are known to be material transducers. The information from these sensors is transmitted to the brain in the form of material neural signals. If the mind were immaterial, then the information carried by these neural signals would have to be transformed into the immaterial form of the mind. The immaterial mind would also have to be able to cause material consequences via muscles. It is known that the brain controls muscles via material neural signals; thus the command information created by the immaterial mind would have to be converted into these material signals. However, a converter that transforms immaterial information into material processes is impossible, as energy and/or matter would have to be created from nothing.

Cartesian dualism has also another difficult problem that actually makes the whole idea unscientific. The assumed immaterial mind would power cognition and consciousness, and would be the agent that does the thinking and wanting. But, how would it do it? Is there inside the brain some immaterial entity, a homunculus (small person) that does the thinking and perceives what the senses provide to it? How would this entity do it? Immaterial entities are, by definition, beyond the inspection powers of material means, that is, the material instruments we have or might have in the future. Therefore we cannot get any objective information about the assumed immaterial workings of the mind; any information that we might believe to have via introspection would be as good as any imagination. (Psychology does not tell anything about the actual enabling mechanisms of the mind. In the same way, you will not learn anything about the innards of a computer by inspecting the outward appearance and behavior of the executed programs.) Thus, the nature of the immaterial mind would remain in the twilight zones of imagination and make-believe, outside the realm of sound empirically testable science. By resorting to the immaterial mind, Cartesian dualism has explained what is to be explained by unexplainable; this is not science, this is not sound reasoning, this is not reasoning at all.

The conclusion: Cartesian dualism is wrong and immaterial minds do not exist. The mind appears as immaterial only because we do not perceive the underlying material machinery, the neurons and their firings. The brain does not observe the neural firing patterns, and consequently does not determine that this and that firing pattern would represent this and that entity. The brain does not inspect its neurons and their firings as such and does not naturally create mental images of these. (Now that we know that neurons exist, we can imagine them, but this is not the point.)

4.3. Panpsychism

Cartesian dualism presents that consciousness is a property of an immaterial soul. If this is accepted, a question arises: What is this soul and where does it come from? Panpsychism proposes an answer.

Panpsychism is a dualistic hypothesis that explains consciousness with an assumed immaterial substance, panpsyche. This is a universal, primordial consciousness that is present everywhere, making all things conscious to some degree. Thus, also human consciousness is nothing more than a limited, local manifestation of panspyche. The philosophical roots of panpsychism go back thousands of years, and many famous philosophers have presented their own versions of this hypothesis.

Like many other hypotheses on consciousness, panpsychism may appear at first sight a valid theory that really explains the mystery of consciousness for good. Unfortunately it does not. A scientific explanation tries to explain more with less. Panpsychism does the opposite; it explains the human consciousness with something more complex that cannot be explained at all, something that is a figment of imagination. In doing so, it does not explain anything, and fails to be a scientific explanation. Panpsychism does not save Cartesian dualism, but would it be possible to save the dualistic view by some other hypotheses?

4.4. Property Dualism

No matter how wrong Cartesian dualism would seem to be, we may still feel that there must be a fundamental difference between mind and body, between subjective experience and neural processes. Could we somehow save the mind-body division while rejecting the immaterial substance? *Property dualism* tries to dot that.

Property dualism is a materialist theory that rejects the existence of an immaterial soul. There are no separate substances for the mind and body. The only substance is the physical material one with its material events. However, according to property dualism, certain material assemblies and events have two different kinds of properties, namely physical ones and mental ones. The mental properties emerge from the events of the physical matter, but *are not* reducible to those. According to property dualism, mental events supervene on material events, they "appear on top of these". By definition, a *supervenient* phenomenon S emerges from the material event M in such a way that whenever S is present, M must be present, too. Furthermore, S can change only if M changes. However, S does not have to be present every time M is present. In this respect, a supervenient phenomenon S can be compared to a shadow. S is like the shadow of M, present only when there is light and M is present, but absent, when M is present in darkness.

What would be the mechanism of emergence here? Property dualism does not have a good answer, because it is maintained that emergent mental properties are not reducible to physical properties. Therefore, from the engineering point of view, emergence is not an explanation at all. At best it is a quasi-explanation, where an unexplainable mechanism, the emergence, is used to explain the original problem. Therefore it is difficult to see, how property dualism could really explain anything.

4.5. The Identity Theory

The identity theory solves the problem of emergence by denying supervenient phenomena. According to the identity theory, mind is the action of nerve cells and their connections. Francis Crick summarises this in his book *Astonishing Hypothesis*: "Your sense of personal identity and free will are in fact no more than the behavior of a vast assembly of nerve cells" [Crick 1994].

There are some varieties of the identity theory. The *type-type identity* model proposes that a certain type of neural activity always corresponds to a certain type of mental activity and vice versa. The *token-token identity* model proposes that a certain type of mental activity may be generated by different types of neural activities. This would be just like in the computer, where, for instance, a certain kind of graphical interface may be generated by different algorithms. The token-token model should appeal to engineers as it would seem to allow the realization of the neural mechanisms by various ways, also by non-biological electronic ways.

The identity theory is a simple one, apparently working without any hypotheses about emergence or other quasi-explanations. As such, it would seem to be the one to be preferred. However, difficult problems remain. If the mind is the same as the neural processes of the brain, then why are some neural processes perceived as conscious mental content, while others are not? Why is no mental content perceived during dreamless sleep, while many neural processes still continue in the brain?

The identity theory does recognize the above problem and proposes an answer: Those neural states that are identical with conscious mental states, have an additional layer of mental properties. There you have it, but somehow this proposition seems to weaken the whole point. It seems that instead of explaining the mind in terms of neural states, the identity theory is introducing another layer to be explained, and in this way the searched explanation slips further away. This problem has been observed by some philosophers. For instance, Maslin laments that the identity theory may escape the dualism of substances (with a very narrow margin perhaps), but not the dualism of properties [Maslin 2001]. This takes us back to the property dualism and, by a nasty surprise, the concept of emergence re-emerges.

It is an empirical fact that neural processes exist and some of them correlate with consciously perceived mental content, and therefore there may be some truth in the identity theory. It is also obvious that neural processes may appear as conscious mental content only under certain conditions, but what would these conditions be? Property dualism and identity theories have not provided working answers. Would the investigation of the correlation between conscious mental content and neural processes provide answers?

4.6. Neural Correlate Theories

Neural correlate theories are related to the Identity theory. It is known that in the brain there are material neurons, synapses and glia, where physical processes take place, and without these there will be no consciousness, no subjective experience. Nowadays these material neural processes can be externally detected by a number of methods and instruments, and by these means some correlations between neural processes and reported subjective experiences can be found, see e.g. Reggia *et al.* [2016]. Therefore, should we just look for the neural areas in the brain that are correlated with conscious experience, and when these areas, the neural correlates of consciousness (NCC), are found, declare the phenomenon of consciousness solved for good in the best spirit of a modernized identity theory?

Some eminent researchers like Crick and Koch [1992] have investigated along these avenues, This work is important and necessary, because the workings of the neural machinery must be studied and understood before we can fully explain the workings of the biological brain. But, would this research explain also the appearance of the conscious mind? For instance, it may be observed that the activation of a certain group of neurons at the visual cortex is associated with the subjective experience of red. But, does this observation really explain, how the experience of red would arise? Not really. Also, it is known that certain opioid peptides like endorphin affect synaptic transmission and cause euphoric effects. Is this the explanation for the experience of feeling good? No, not really. The crucial step in the explanation remains missing; how the releasing of the opioids in the brain generates the subjective feeling of pleasure. There is nothing in these proposed explanations that would explain, why and how any subjective experience would arise from these processes. NCC theories may look shiny and modern, but their explanatory power is not any better than that of those old philosophical theories.

Something is missing. Obviously the missed point is not to be found in neurons as such. Perhaps it could be found in the transactions between neurons, also known as information integration?

4.7. Information Integration Theories

Some researchers have proposed that conscious states are related to the active cross-connections between the various parts of the brain and also the author has recognized this connection. Therefore, would information integration alone be able to generate consciousness?

There are some attempted explanations of consciousness by information integration. The Information Integration Theory (IIT) of Tononi proposes that consciousness arises from the brain's (or other cognitive system's) capacity to integrate information [Tononi 2004]. Balduzzi and Tononi [2009] have proposed that the quantity of consciousness is the amount of integrated information, and qualia are specified by the set of informational relationships between the elements of the integrated information. Balduzzi and Tononi try to describe geometrically the entire set of informational relationships of the integrated information by considering a "qualia space" with an axis for each possible quality. In this way each quale can, in principle, be mapped geometrically, and a corresponding neural activity pattern may, perhaps, be found. However, this approach does not explain how the neural activity would give rise to any inner appearance of qualia in the first place; the actual problem of qualia is neither touched nor solved. Therefore, the problem of consciousness remains unsolved, too. Brooks [2014] notes that "IIT is barely on its feet" and does not explain what is going on in the individual structures of the physical brain.

Von der Malsburg [1997] has also recognized the role of cross-connections and integrated activity in consciousness. Also experimental evidence seems to point towards this conclusion. For instance, the research of Massimi *et al.* [2005] seems to show that the active neural communication between the different parts of the cerebral cortex were related to consciousness. During dreamless sleep this communication is subdued and consciousness has vanished.

Likewise, the function of information integration via the focusing of attention on the same topic by "broadcasts" is an essential feature in the cognitive models of Baars [1988] and Shanahan [2010]. In their models this activity is the one that is said to make the distinction between conscious and non-conscious processing.

As stated before, it is obvious that active cross-connections between the various parts of the brain (or an artificial cognitive system) are necessary for the realization of one feature of consciousness, namely that of reportability; conscious states are reportable states, to inside and to outside. Without the cross-connections and the exchanges of broadcasts between the various brain areas no reports can be transmitted across the brain and no associative memories of the event can be created. Without these cross-connections and internal reports we could not even reach out and touch objects that we see. An event that does not have any effects and cannot be remembered even for a little while, is hardly recognized as a consciously perceived one.

But, make no mistake here. The mere number of active connections or the quantity of "information integration", no matter how large, does not make the system conscious. The primary criterion for consciousness is the requirement of the internal appearance in the form of qualia. The requirement of information integration is secondary. Cross-connections and information integration facilitate reportability, memory making, situation awareness and many cognitive functions as well. However, information integration alone does not explain consciousness, because it does not address the explanatory gap of qualia; it is only related to some secondary aspects of consciousness.

4.8. Dissecting Consciousness

A quick research into the consciousness literature reveals a conceptual problem; "consciousness" is used to refer to different aspects of mind and cognition. Sometimes "consciousness" is used to refer the totality of the conscious experience, but this leads to difficulties. What exactly should be explained? If you do not know, what exactly is to be explained, then most probably you will explain something else, or maybe nothing at all. Therefore, it helps to use the well-known problem solving strategy of dissecting the problem into well-defined sub-parts.

Block [1995] has tried to do this by dividing the concept of consciousness into two components, which he calls the phenomenal (P) consciousness and the access (A) consciousness. P-consciousness relates to the raw perception of external and internal sensory information and feelings, while A-consciousness relates to the ability to reason, remember, report and control behavior. However, A-consciousness is not really consciousness at all, as the mechanisms behind recall and thinking are subconscious; only the results of these are consciously perceived.

It is obvious that A-consciousness may be explained by the operation of information processing mechanisms. On the other hand, P-consciousness relates to the phenomenal aspect of consciousness, "how everything is experienced". Explaining P-consciousness would involve the explanation of phenomenality; what are qualia, what is feeling, why it feels like something, and how a cognitive system can feel anything. Cognitive abilities may be externally assessed, but the subjective phenomenal experience remains personal and can be observed only indirectly.

A-consciousness has function, as it is related to the cognitive abilities of the subject. P-consciousness may or may not have a function depending on its definition by different researchers. In order to clarify this situation, Boltuc [2009] has proposed an extended division of the concept of consciousness. According to Boltuc, Block's P-consciousness can be divided into functional first-person consciousness and pure subjective phenomenal consciousness, which Boltuc calls H-consciousness. Boltuc's functionless H-consciousness is the primary subjective experience, without which there is no consciousness. Thus, obviously H-consciousness relates to the hard problem of consciousness and is the phenomenon that should be explained in the first place.

Some AI researchers have proposed that in robots the hard problem of consciousness can be by-passed by the realization of functional consciousness, that is, by the realization of the functions of consciousness without any phenomenal internal experience. How is this done? For example, when you put your finger in the candle flame, you will experience pain and will withdraw your hand quickly maybe with some verbal reports. The experience of pain is the phenomenal aspect, and the withdrawal with some carefully selected comments is the functional reaction. Pain will also teach us not to do again things that lead to pain, not to put our finger in the candle flame. Similar learning can be realized in a machine without any real *feel* of pain. *Artificial functional pain* (like in a trivial machine or a computer simulation) can be defined as the process that realizes some or all functional aspects of pain. In a similar way, *artificial functional consciousness* could be defined as a process that realizes the functions of consciousness without the phenomenal experience.

But, are there any functions that consciousness would execute? Does consciousness think, reason or plan? No. Phenomenal consciousness is only the internal appearance of the mental content. Thinking, reasoning and planning belong to cognition, which can be explained without any reference to consciousness. Thus no cognitive functionality remains for consciousness.

However, consciousness does have one function; namely the act of the grounding of symbol meaning via self-explanatory information. This function cannot be executed without qualia, because qualia are the form of self-explanatory information. If this function of consciousness were implemented in a system, then the system would have phenomenal experience and be conscious. Consciousness does not have any non-phenomenal functions and therefore "functional consciousness" does not exist. Consequently, artefacts that are claimed to be conscious, because they implement "functional consciousness", are not conscious at all.

Harnad and Scherzer [2007] propose that consciousness is phenomenal. According to them, to be conscious is to have phenomenal feel; consciousness is feeling, nothing more, nothing less. This feel has no causal powers. The difference between conscious and non-conscious acts is feeling; conscious acts are felt, while non-conscious acts are not. According to Harnad, every conscious percept involves feeling; a conscious subject feels pain, fear, hunger, etc. Also visual perception would involve feeling, it would feel like something to see blue.

This view can be challenged. For instance, the author finds visual perception totally neutral and without a slightest feel comparable to the actual feel of something, like a touch, temperature, pain or other bodily feel, yet the author finds himself to be visually conscious. It may be argued that the phenomenal qualities of visual perception (colors etc.) would constitute the feel, but this would stretch, twist and redefine the meaning of "feel" towards the point, where "feeling" and "consciousness" were synonyms; "consciousness is feeling". At that point the explanatory power of the concept "feeling" would be lost, because synonyms do not explain anything. For instance, if you do not know what "a car" is, and you are told that it is "an automobile", you will not be any wiser. Therefore "feeling" in its conventional meaning is not a sufficient constituent of consciousness. The spectrum of the phenomenal experience of consciousness involves more than feeling.

Scientific explanations are causal explanations; they propose mechanisms that cause the phenomenon that is to be explained. Harnad and others have proposed that real phenomenal consciousness does not have any function, and has therefore no causal powers. Thus consciousness is not an executive agent. All the functions that consciousness might seem to have, are effected by the neural processes behind the conscious experience, not by the conscious experience itself. Does this lead to the conclusion that consciousness cannot be causally explained at all? At least Harnad [Harnad and Scherzer 2007] would seem to think so. According to Harnad, natural forces are causal, but without any feel. Therefore the feel of consciousness cannot be explained with a combination of natural forces. Harnad posits that causal force explanation would call for a natural mental force that has feel, such as a telekinetic force. However, mental telekinetic forces with or without feel do not exist, and therefore a causal explanation for the feel of consciousness cannot be provided in this way.

It is true that physics equations do not yield the feeling of feel from the application of natural forces. And how could there be; the equations of physics describe numerically quantified entities and produce numerical values, not any qualities of phenomenal experiences. No matter how you manipulate the equations, no feel will turn out. This is a wrong way to look for phenomenal feelings. For the explanation of feel you must study self-observing systems that have subjective experience.

For survival, a conscious being must have various cognitive skills, but for being truly conscious, it has to have some kind of a phenomenal subjective experience. The ultimate explanation of consciousness is not an explanation of cognitive skills, it is not an explanation of any "A-consciousness". Instead, it must focus on the hard problem of phenomenality and the explanatory gap between qualia and material processes. This is done in the following chapter.

Chapter 5

The Explanation of Consciousness

5.1. The Real Problem of Consciousness

Introspection shows that the contents of our moment-to-moment consciousness consists only of sensory percepts and percepts of our thoughts and emotions. This leads to the idea that consciousness is related to perception. The idea of consciousness as perception is not a new one, it has been proposed already by David Hume (1711–1776): "I can observe nothing but perception" [Hume 2000].

Introspection also shows that all our conscious percepts have the form of qualia, phenomenal experience. Therefore one could infer that consciousness is the presence of inner phenomenal subjective experience. This idea has been shared, for instance, by Gamez [2008], who has stated: "Consciousness is the presence of a phenomenal world". This proposition is also seconded by Aleksander [2009]. Also Harnad has emphasized the phenomenal aspect: "Consciousness is feeling" [Harnad and Scherzer 2007]. These ideas do not constitute an ultimate explanation of consciousness, but they point towards the real problem of consciousness: Consciousness is phenomenal, but how can a material brain produce the inner impression of phenomenal experience?

It is generally understood that at least in principle, all physical processes can be detected and measured by physical instruments via physical interactions between the detector and the detected. Accordingly, various physical brain imaging methods are able to detect neural activity patterns and neural signals in the brain. However, no instrument has ever been able to detect qualia and phenomenal inner experience. Pain carrying neural signals can be detected, but the actual feel of pain remains undetected. The same goes for qualia in general. Phenomenal experiences cannot be detected by physical instruments. Does this show that qualia and consciousness are non-physical, immaterial entities, and dualistic explanations of consciousness have to be valid after all? On the other hand, if it could be shown that qualia were not immaterial, dualistic explanations of consciousness would be false and useless.

The question is: Why and how does some of the neural activity in the brain manifest itself as apparently immaterial qualia instead of the actual material, neural activity as such, or not at all. This question is known as "the explanatory gap" [Levine 1983], and also as "the hard problem of consciousness" as recognized by Chalmers [1995a, 1995b]. The solving of this problem would constitute the explanation of phenomenal consciousness. The issues that relate to the contents of consciousness, such as self-consciousness, situational awareness, social consciousness etc. are consequential and do not have a part in the explanation of the basic phenomenal consciousness.

The complete problem of consciousness includes the following questions:

1. How does some neural activity appear internally as the subjective experience of apparently immaterial qualia?
2. Introspection: How can mental content be consciously perceived?

The solving of these problems constitutes the explanation of consciousness. The first question is the crucial one; any proper theory of consciousness must address it, otherwise the theory does not explain consciousness. The theory that solves these problems should be a practical one that would allow, at least in principle, the creation of conscious robots and artefacts; if we cannot reproduce it, we have not truly understood it. Thus, the explanation of consciousness should not be a dualist theory, and it should not rely on the concepts of emergence and supervenience.

In the following an explanation of consciousness is presented, one that rejects material and immaterial substances, dualism, panpsychism, emergence and metaphysics generally. What remains is obvious and allows also the artificial creation of conscious machines.

5.2. The Explanation of Qualia and Phenomenal Experience

The phenomenal experience of being conscious is subjective and accessible only by introspection. Introspection shows that the contents of consciousness consist of percepts with the form of qualia. Qualia and consciousness are linked together. When qualia cease, percepts cease and the contents of consciousness becomes zero. The subject will not be aware of anything and, by definition, will be unconscious. Thus, to *be conscious is to have phenomenal experience with qualia,* therefore the explanation of qualia and phenomenal experience will also explain consciousness.

On the other hand, qualia are self-explanatory information, necessary for the grounding of meaning of symbols. Thus the explanation of qualia is also related to the solution of the symbol grounding problem.

Qualia has a fundamental problem: The brain's perception processes operate with neural signals and signal patterns. However, this is not our subjective experience. Instead of perceiving the actual neural activity, we perceive the external world, body, and our mental content in the form of qualia, and this is the form of our contents of consciousness. The neural processes and neural signals can be traced and observed by physical means and instruments, but the inner appearance, qualia, has remained inaccessible as if it were immaterial.

The fundamental problem of qualia and phenomenal experience to be explained is twofold:

1. Why do qualia appear as immaterial?
2. How can some neural activity appear internally as qualia, instead of appearing as what it is, or not appearing at all?

This twofold problem is the essence of the hard problem of consciousness and also the explanatory gap. It will be solved here.

Firstly, why do qualia appear immaterial, not observable by physical means? This is an important question, because it might indicate that qualia are not physical. If that were the case, then how could qualia function within the material processes of the brain? Dualism and the mind-body problem would be unavoidable.

Modern brain imaging instruments are able to detect neural activity in many ways, and correlations between mental content and the neural activity have been found. Why then these instruments have not been able to capture the actual internal appearance of qualia? Would it be possible that in the future more advanced instruments could be able to capture the internal appearance and solve the problem of qualia and consciousness in this way? (See Fig. 5.1.)

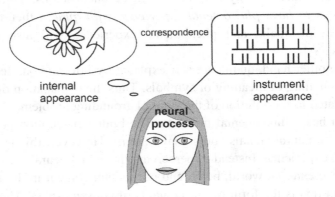

Fig. 5.1. Neural processes may be inspected by various instruments that produce instrument-specific data, "instrument appearances". Some, but not all, neural processes in the brain have an internal appearance; they depict something and the subject experiences these appearances consciously.

In order to understand the situation, *the fundamental limitations of measuring* must be considered. Measuring instruments and arrangements detect and measure only the property that they are designed to measure. Measurement is an interaction between the instrument and the measured quantity, and the result is an appearance provided by the instrument, it is not the measured entity. For example, if you measure a photon as a particle, the photon will appear as a "particle". If you measure a photon as a wave, the photon will appear as a "wave". However, the particle view and the wave view are only our own *models and descriptions* of the photon, while the photon as itself is what it is. Measurements do not reveal the real essence of the photon, the Kantian "das Ding an sich". The same goes for all measurements; the measured object is not revealed as itself, instead our instruments give some symbolic patterns and numeric values that represent and describe some properties of the measured object. Instead of capturing real essences as such, only descriptions of these are produced. Thus, the failure to detect and measure qualia as such is the direct consequence of the universal limitations of detection and measurement processes; it is not possible to access externally the detected entity as the phenomenal itself. The only instrument that can detect phenomenal qualia as such is the experiencing system itself, because qualia are part of the system itself, das Ding an sich. Therefore, the non-detectability of qualia is not an indication of any non-physical miraculous nature of the same, and it cannot be used as a proof of any immaterial mind or soul, or as a proof that qualia cannot be artificially created. The non-detectability of qualia is not a proof against conscious machines. Qualia only appear as immaterial, because they cannot be physically captured due to the general limitations of measuring.

The second part of the hard problem of consciousness is about the relationship between neural activity and qualia: How can some neural activity appear as qualia, instead of appearing as what it is, or not appearing at all?

In perception, external entities are inspected by the sensors. Sensors generate static and dynamic neural signal patterns as reactions to the sensed stimuli, and these patterns convey the sensory information to the brain. Now here is the tricky part: Why does the impression of sensed qualities arise, instead of the impression of patterns of neural signal firings? How can physical, neural circuitry carry impressions of sensed qualities, and yet remain hidden or transparent?

This situation can be illuminated with a simple experiment. Take a rigid stick, maybe a pen, in your hand. Next, scan a rough surface with the tip of the stick. What is your experience? Instead of sensing the vibrations of the stick against your fingertips, you get the impression of the roughness of the surface. The apparent origination point of the roughness percept has been *externalized*; it is not in your brain, it is not in your fingertips, it is on the rough surface, because it is the surface that you have inspected. The vibrations of the stick have carried the information about the surface, while the stick itself has remained transparent. The actual constitution of the rigid stick, may it be wood, metal, etc. has no relevance here.

In a similar manner, sensory neural signal patterns carry the sensed information, while the carrying machinery and signals as themselves remain hidden. The neurally carried spatial and temporal patterns reflect the qualities and properties of the sensed entities, and appear as self-explanatory qualia; apparent qualities of the world. This neural activity does not appear as itself, because it is not inspected as itself, and besides, in the brain there are no sensors or processes for that purpose.

The sensory neural patterns appear as the carried information, because they are about the sensed stimuli. These patterns are also subject to the general limitations of measuring. They are not the real sensed phenomena, but operate as "false color" impressions of the world, however taken at their face value.

The answer to the question "how can some neural activity appear as qualia, instead of appearing as what it is" will be: We are not able to inspect and observe any neural activity in the brain as a biological nervous process. Therefore, if a neural activity becomes internally observed, it will not be observed as such. Instead, it will be observed as the carried information, appearing as externalized qualia, properties of the body and world, just like in the rigid stick example. The neural machinery is only a transparent carrier for information.

One question remains: What is it exactly that makes qualia consciously perceived? This question can be answered by inspecting the difference between conscious and non-conscious perception.

5.3. The Difference between Conscious and Non-Conscious Perception

In the foregoing it was stated that qualia is the name for the qualitative ways in which information manifests itself in the mind of a conscious subject, and to be conscious is to have qualia-based inner subjective experience. Obviously, to have qualia is to be aware of them. For example, you are not in pain, if you are not aware of it. You do not taste the sweetness of sugar, if you do not experience it consciously. To have qualia is to experience, and therefore qualia would seem to be inherently conscious, but what exactly does cause that?

The simple answer is reportability, the ability to remember a percept long enough so that a report can be generated. This report can be in many forms. If the subject is not able to report a percept to itself (at least), then obviously the percept has been sub-conscious, or non-existent. For instance, the reporting of a sensed sound may include the following.

- I can hear it and I can focus my attention on it (qualia and attention)
- I can report that I hear it (report)
- I can remember hearing it (memory making)
- I can turn my head towards its source (sensorimotor information integration)
- I can associate some meaning with it (information integration; learning and evocation of associated meaning; the sound of a bell, etc.)
- I can report the associated meaning

If a sound is only heard fleetingly, no report can be generated to oneself, and nothing of the above can take place; the perception of that sound has remained sub-conscious, but it may still have some effects.

Fully conscious perception leads to a number of consequences that are facilitated by a large number of cross-connections; the activation of associative links and the evocation of responses. This takes some time, and it is known that the brain requires a certain minimum duration for a sensory stimulus in order to perceive it consciously. If a sensory stimulus is too short, it will not capture wide attention. It is not able to activate associative links before dying out, and therefore only limited and short-lived memory traces will be generated, evoked responses will be minimal or non-existent, and no report can be given. There will be minimal or non-existent system reactions. Without system reactions there will be no such qualia as pain and pleasure.

The above observations apply to conscious perception in general. The difference between conscious and non-conscious perception is the difference between information integration in each case; the quantity of activated connections that enable reporting, recalling, evoking of associated meanings and response generation.

Regardless of the number of evoked connections, the basic essence of consciousness is the inner manifestation of sensory percept neural patterns in the form of qualia. In this way external sensed entities can be consciously perceived, but how about the perception of mental content, as there are no internal sensors in the brain?

5.4. Conscious Perception of Mental Content

In the above it was stated that consciousness is based on sensory perception. How then does this dovetail with the fact that we are not only aware of the sensed world, but also of our thoughts and imaginations? How can these be perceived in the proposed way, as obviously the neural activity behind these is located deep in the brain, beyond the reach of sensory perception?

The phenomenon of inner speech gives some clues. Our verbal thoughts have the form of inner speech, as if we were silently talking to ourselves. We are the talker and the listener, and therefore we should already know, what we are going to say. This is not the case, and it is evident that our thoughts arise from the brain's sub-conscious neural processes, which are not consciously observable. Therefore the products of these processes must be translated into a perceivable form, and that would be overt, spoken speech, which could be perceived by hearing.

However, speaking one's thoughts aloud all time would reveal one's thoughts to everybody, and that would not be very nice. Therefore, instead of speaking aloud, it would be better to route the thoughts internally, in the form of neural signals, to the auditory perception module. But how could this happen?

Cochlear implant technology shows that the stimulation of auditory sensory nerves will lead to the internal percept of the sound qualia that would normally result from the operation of the outer ear and cochlea. Thus it would be technically possible that in the brain this is what happens; inner speech-related neural signals are routed back to the auditory sensory cortex, where they would internally activate corresponding auditory sensory neural signals. This would create the impression of heard speech. This leads to the principle of feedback.

Feedback loop models [e.g. Chella 2008, Haikonen 1999, 2003, 2007, Hesslow 2002] propose feedback loops that bring back cortical output signals to the input neurons of the sensory modalities. Here these feedback signals would excite sensory feature signal patterns. The feedback information would be transformed into virtual percepts, and in this way the brain could introspect its mental content by using its perception processes, see Fig. 5.2. Also some studies seem to show that inner speech does indeed utilize auditory perception brain areas, especially those that are mostly used for processing external speech [Scott 2013].

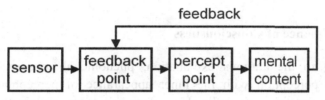

Fig. 5.2. The perception of mental content by internal feedback. Te feedback loop feeds some mental content back to the perception process, and this content will appear as a "virtual percept" at the percept point, and will be handled, distributed and remembered like a real percept.

With the internal feedback linguistic thoughts could be consciously perceived as virtual percepts of silent inner speech. Inner speech would not necessarily have all the tonal qualities of actually heard speech. It would also lack the direction quality, and therefore its perceived location of origin would be inside the head.

The internal feedback principle is applicable to all sensory modalities, and instead of one feedback loop there would be a great number of them. When we introspect our mental content, such as inner speech, inner imagery and the like, we will notice that all these are percepts and appear to us as qualia of their own kind. Inner speech appears as kind of heard speech, our imaginations appear as vaguely "seen" patterns, imagined body movements appear as virtually executed movements, etc. Feedback generates qualia to mental content. Having reportable qualia is to be aware of them, and in this way each sensory modality is reused for the conscious perception of mental contents.

It is possible that the developing brain of a child does not initially have these feedback connections. In that case the internal feedbacks would have to be learned in early childhood, and before this the only way in which a child could perceive inner speech would be to talk aloud. In this way the feedback loop would be learned and rehearsed. This would explain why little children must talk aloud all the time, a phenomenon that surprises many new parents. Likewise, the development of imagination would call for corresponding playful exercise in early childhood; drawing, manipulation of objects etc., actions that would build internal feedback loops in visual and motor modalities.

5.5. The Essence of Consciousness

According to the author's explanation, phenomenal consciousness is reportable perception with the inner appearance of perceptual neural patterns in the form of qualia. How does this view explain the actual essence of consciousness?

The apparent conclusion is that if consciousness is not an immaterial substance, then it must be a material one, no other possibilities exist. Surely nothing can be left, if both materialism and immaterialism are rejected. But is it so?

Consider a printed book; where is the story and information? It is not in the atomic composition of the ink, instead it is in the ink patterns that form the letters and words. Likewise, where is the music in a vinyl record? It is not in the actual chemical composition of the vinyl, it is in the varying form of the groove. The study of the atomic composition of the carrier media and matter will not lead anywhere. The information is the forms and patterns, not in the carrying substance.

In the brain the situation is the same, with a small difference. In books and vinyl records the patterns are static, but in the brain they are time-varying and interactive. These patterns carry the information and meaning, while their material neural platform is only an enabler.

Perception and thinking are interactions between neural activity patterns. The internal appearance of these patterns is the appearance of conscious mental content, and that, like the appearance of the printed text of this book, is not a material or immaterial substance. Yet, right now this text is causing neural activity in your brain. The internal appearance of qualia carries information and has effects, but it does not have executive powers; it is neither a self nor an agent. It just an appearance; not matter, not an immaterial substance.

The final point is here: Traditional attempted explanations of consciousness fall victim to category error. Consciousness has been seen as material or immaterial substance, agent, entity or soul. It is none of these, and therefore the search for consciousness within those categories has failed and will fail. Consciousness is neither matter nor an immaterial entity, and the mind-body problem does not exist in the first place.

Consciousness is just a way of operation; one with qualia as self-explanatory information. Philosophers have tried to explain something that just is not there, and with wrong mental models for that matter.

However, this raises profound questions. If there is no soul, then what are we, what is the wanting and decision-making self? What is self-consciousness? How does the impression of I arise?

5.6. The Explanation of Self-Consciousness

Self-consciousness is the subject's ability to perceive its own existence as a discrete being, separate from the environment. Auditory and visual percepts are externalized, the environment with its things is out there, and the observing self is a kind of vantage point in the middle of everything. Self-consciousness builds upon the phenomenal consciousness, and it is based on the perception of the properties of the subject's body, its conditions as well as the subject's mental content; thoughts, imaginations and emotions.

A self-conscious subject has a body image and a mental self-image. Self-consciousness includes also the concept of *first person ownership*; my body is mine, my thoughts are mine, my memories are mine, my decisions are mine, my speech is mine, my acts are mine.

The percepts of the body give rise to the *body image* that relates to the shape, size and function of the body, and the difference between the body and the environment. In humans and animals the border between self and the environment is usually clear. The body belongs to the subject, while the environment does not. The body goes where we go, while the solid environment does not. The body is neurally wired to the brain, while the environment is not.

The neurally wired network of various body sensors is called the somatosensory system. This system consists of several sensory modalities with a large number of receptors that sense body part position (proprioception), touch (tactition), pain (nociception) and temperature. The somatosensory system allows the brain to monitor the status of the body continuously. The information delivered by the somatosensory system leads to two simple rules: If it hurts, it is a part of mine, not to be eaten by me. If I can feel it, it is mine.

The first rule is seemingly rather ridiculous, but it is not, as witnessed by the unfortunate sufferers of the congenital insensitivity to pain syndrome (CIPA). These subjects do not feel pain, and may actually try to eat their tongues and lips, etc. They will not readily notice when they hurt themselves, and will not automatically learn to avoid dangerous activities. CIPA sufferers usually die rather young.

The second rule is manifested, for instance, during an occasional numbness (paresthesia) of a hand due to a bad sleeping position. It may then happen that for a scary moment the subject recognizes the hand as a foreign object.

The *mental self-image* relates to the ideas that the subject has about itself; who I am, what I feel, what I want, what I am able to do, how I am in respect to others. The mental self-image calls for the retaining of personal history in the form of memories.

The group of ideas, which contains a body image and a mental self-image that are grounded to the material self via the somatosensory system, can be called the subject's *self-concept*. A self-conscious subject can refer to itself in many ways within the self-concept group of ideas.

Based on the foregoing, the prerequisites for self-consciousness can be listed. In addition to the general requirements for phenomenal consciousness, a self-conscious agent should have a somatosensory system, body image, mental self-image, self-concept, concept of first person ownership and remembered personal history.

5.7. Implications to AI and Machine Consciousness

Without understanding there cannot be any real intelligence. Without the acquisition of meanings there cannot be any understanding. Basic meanings cannot be acquired in symbolic form, because symbols call for interpretation; the meaning must come from elsewhere. Meaning has to be associated with its symbol, but initially this meaning cannot be in the form of another symbol demanding another explanation. This leads to the symbol grounding problem, which cannot be solved within pure symbol systems. Therefore meanings must be imported in forms of self-explanatory information. Self-explanatory information has the form of qualia, and to have reportable qualia is to be conscious. Future AI systems that really understand, would have to be conscious.

The explanation of consciousness given here is not a metaphysical one, and it does not posit any immaterial entities or substances. On the contrary, it points out that in the brain the carrier of the contents of consciousness is not observed, and the material constitution of the carrier is not included in the equation. Thinking is manipulation of content patterns, not the material constitution of carrier media. Therefore, instead of biological neurons and the other stuff, it should be possible to use also artificial carriers, like electronics, as long as the basic requirements for consciousness are met.

From the given explanation the basic requirements for consciousness are 1) Sensory perception process, which produces self-explanatory sensory percepts in the form of qualia, 2) Feedback loops that allow the introspection of the mental content in terms of virtual percepts, 3) Cross-connections and short-term memory for the generation of report.

How could these be technically realized? Among others, James Reggia has noted that it is difficult to understand how higher-level cognitive computations could arise from the brain's lower-level neuro-computational processes [Reggia *et al.*, 2014, 2018]. Reggia calls this problem the *computational explanatory gap* (CEG). The solving of the computational explanatory gap would amount to the solving of the problems of the artificial realization of higher-level cognitive processes, especially with meaning, which calls for self-explanatory information.

Unfortunately, traditional computational AI is not able to utilize non-symbolic self-explanatory information, and therefore it cannot escape the symbol grounding problem and cannot operate with meanings. Thus it will not provide the required solution.

Would artificial neural networks do better? Traditional artificial neural networks and algorithms, such as Back Propagation, Deep Learning and Self-Organizing Maps, execute statistical computations, and can be used as pattern recognizers and classifiers. However, thinking and cognition are much more than statistical pattern classification. In addition to sub-symbolic information processing, cognition operates in symbolic ways, too. Traditional artificial neural networks are not well suited for this.

Would the combination of digital computers and artificial neural networks be the solution? These kinds of hybrid solutions have been tried with limited narrow area success, but real Artificial General Intelligence, AGI, has been elusive. So far no combination of existing technologies has produced true cognition, intelligence or consciousness.

Is the computational explanatory gap unsolvable, then? Yes, it is, and there is a fundamental reason. Symbolic calculations and algorithms cannot give rise to phenomenal experience in the way of human cognition and consciousness. Computations are not and cannot be accompanied by feel and meaning to the computer itself.

Machine consciousness requires a different non-computational approach. It also calls for the realization of the various processes of cognition with meaning, from perception to thinking and response generation. Biological inspiration helps here. For the design of a conscious robot, non-computational artificial neurons and networks must be devised with an architecture that is able to model and produce the necessary higher level cognitive functions.

This is done in the following, and it will lead to an associative neural architecture that is able to support and produce conscious information processing. Consciousness is based on perception, therefore sensory perception that produces self-explanatory information for the grounding of meaning is treated first.

Chapter 6

The Gateway to Mind; Sensory Perception

6.1. The Form of Sensory Percepts

Senses are the gateway to the mind. All information about the environment and the body comes via sensors as direct percepts. Also indirect information in the forms of linguistic, graphic or other descriptions comes via sensors and is carried by direct percepts. All consciously sensed information appears internally as qualia. Each experienced quale has its own self-explanatory meaning, namely the actual or virtual presence of the physical feature that produces the quale.

Sensed natural objects and entities have combinations of details and features that can be inspected separately. These appear as qualia and are carried by sensory neural signal patterns. An object can be recognized as a familiar one, if its appearance, the combination of features, has been learned. Direct learning calls for the detection and memorizing of repeating sensory neural signal patterns that carry the relevant combination of features. Learning by description is also possible, if the relevant qualia and their names are already known.

However, there is a problem. The more complex a sensed object is, the more unlikely its appearance will repeat exactly in the same form, and consequently, the more random it will appear. For instance, an object is rarely seen exactly in the same way due to different viewing angles, distances and illumination, yet it should be recognized as the same, see Fig. 6.1.

This applies to heard sounds, too. A spoken word may sound different, when spoken by different speakers or even by the same speaker, yet it should be recognized as the intended word. It is not practical to learn each and every possible different appearance of an object, and exact large scale pattern matching will not work very well or at all. What else could be done?

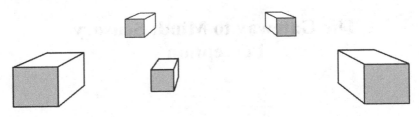

Fig. 6.1. The problem of varying appearance. A box is a box, but it may be big or small, and its appearance may vary. Yet it should be recognized as a box in each case.

The solution is the use of *invariant features*; features that survive transformations. The method known as *distributed representations* is well-suited for this purpose [Hinton, McClelland and Rumelhart 1986]. Each distributed signal or a small-scale signal pattern represent a certain detectable feature, and sensed complex entities can be described by different combinations of these. Entities may now be deemed to belong to a certain class by a small number of invariant elementary features that describe common key details. Human sensory recognition apparently operates in this way, as demonstrated by Fig. 6.2.

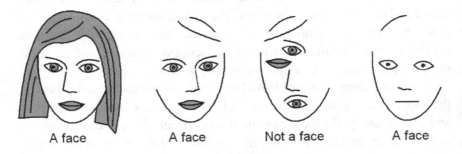

A face   A face   Not a face   A face

Fig. 6.2. Are these faces? No, they are just collections of lines that evoke impressions in the looker's mind. Only some elementary features in proper relative positions are required for the impression of a face. These details are common to all faces.

The strength of the human sensory recognition lies in its imperfection. Seen objects and heard sounds are not recognized via perfect pattern matching. Instead, the perceived combinations of details and patterns only remind of something.

Distributed signal representations are based on signals and signal patterns that stand for a certain feature. Sensors do not necessarily produce these kinds of signals readily, and therefore some kind of preprocessing is required.

For example, a microphone produces a varying electric voltage, which has an analog form of the vibrations of the air pressure. As such it contains the sum effect of the individual sounds that are instantaneously present. For the eventual cognitive perception these individual sounds must be separated from each other by audio spectrum analysis. In computers the spectrum analysis can be performed by filter banks or by Fourier transform.

In hearing the problem is the same; the eardrum vibrates to the sum of all sounds, which must be separated. The inner ear provides the first step towards this by audio spectrum analysis, which is performed by the cochlea. The cochlea produces separate signals for each heard tone from 20 Hz up to 20 kHz, and these carry the auditory quale of each frequency.

It is a mathematical fact that every continuous waveform is the sum of a fundamental frequency and its harmonics. This applies to sounds, too. Based on this, separate sounds can be segregated by grouping their harmonic frequencies together. Also, sound direction and the common onset time of newly appearing frequencies can be used as a binding factor. Without these processes the sum of sounds that enter the ear would appear as cacophony. Spoken words could not be recognized as individual sound patterns, the sounds of musical instruments could not be segregated from each other, and music itself would not exist.

In visual perception visual details must be segregated. These may include contours, brightness, different colors, small-scale patterns and movement. The signals and signal patterns that describe these are the distributed visual feature signals.

A practically invariant set of distributed signals and signal patterns can be taken to represent a given entity. The sensors may, and usually will produce a large variety of additional signals standing for further details, but these are not required for the recognition of the given entity. A face is a face even though the details may differ.

Sensory perception produces also the illusion of externalization; sensed entities have a perceived location outside the brain, even though the actual percept producing neural activity is inside the brain. This most useful effect creates mental visual and auditory landscapes of the environment. These tell what is where, and help to make the distinction between the self and environment. Externalization also makes the distinction between sensed and imagined entities, like heard speech versus one's own inner speech, and seen objects versus imagined objects. Externalization of percepts is discussed in the following.

6.2. Externalization of Sound Percepts

Hearing externalizes heard sounds. Sound waves enter the ears and make the eardrums vibrate, but this is not our experience. We do not perceive nor experience the vibrations of the eardrums, instead we experience sounds coming from different directions outside the head. The sensed sound is externalized, but how is this illusion created?

There is nothing in the auditory sensory signals that would directly reveal the origination point of the information. The neural fibers originate from the sensors, but the carried neural firings do not convey any information about their origination point. The neural fibers are not labeled and neither are the firings; they have only causal connection to the external stimuli. The strength of a stimulus may be encoded into the neural signal, but that does not tell, if the signal is caused by a weak sound nearby or a far away strong sound that appears as a weak signal at that distance, or even that the signal has external origin. Therefore, without any additional information the auditory sensory signal patterns would stand for sounds with undetermined location.

Additional information for the determination of the origin of external stimuli may be gained via explorative acts. In hearing, the simplest act would be the covering of ears, and this would show that the sounds have an external origin.

However, it is not sufficient to experience that a sound comes from outside. Hearing involves also the detection of sound direction, and this calls for additional information processing, which is facilitated by two ears. In the brain, sound direction detection is based on the intensity difference (at higher frequencies) and the time delay difference (at frequencies below 1500 Hz) between the ears and to a very small extent to the directional frequency filtering of the outer ear (pinna). However, the direction information provided by these means is ambivalent, because a sound source in front of or behind the subject would generate a similar percept. The turning of the head resolves this ambivalence.

A simple practical experiment that illustrates the need for auditory experimentation can be executed easily, see Fig. 6.3.

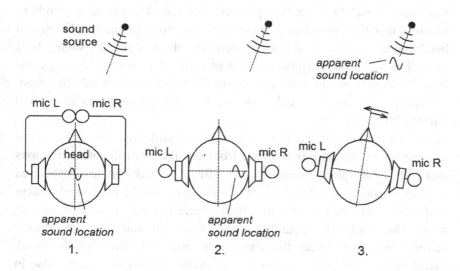

Fig. 6.3. The sound externalization experiment. 1. The left and right microphones are held together creating an apparent sound location inside the head. 2. Microphones near the ears create a transparent system, yet the apparent sound location remains inside the head. 3. The turning of the head resolves the situation and the sound is externalized.

The experiment of Fig. 6.3 should be executed with good quality headphones, a stereo headphone amplifier and a pair of omnidirectional microphones. The headphones must be of the closed design so that no external sound can disturb the experiment. In this experiment the perceived location of heard sounds appears to be initially inside the head, but after an explorative act the sounds become externalized. The sound externalization experiment is executed in three steps:

1. The microphones are placed close to each other in front of the test subject's head. Obviously both microphones receive the same sound intensity and delay, thus the left and right speakers of the headphones generate a sound with same phase and equal intensity. The apparent sound location will be inside, in the middle of the head.

2. The head is kept still and the left and right microphones are brought next to the corresponding headphone speakers. (It is important that the headphones are of the closed design, otherwise acoustic feedback, "howling", can take place!) Now the situation corresponds to binaural listening. Good quality microphones, microphone amplifiers and headphones should become transparent and the test subject should hear as well as when no headphones were present. Thus, it could be expected that the sounds would appear to come from locations outside the head. However, this does not freely happen; the apparent sound locations will still remain inside the head.

3. The test subject turns his head. Suddenly, the sounds are externalized; they appear to come from outside from distinct directions and the headphones seem to go silent. Thereafter the perceived origination points of the sound stay outside even when the head is kept still, and it will be possible to turn the head towards a sound source at will. (The vestibular apparatus of the inner ear and the neck muscle tension receptors detect the changes in the orientation of the head, therefore no visual cues are necessary and this experiment works also in total darkness.) Obviously in the above experiment the pinnae have no influence as they are covered by the headphones.

The lesson from this experiment is: Static perception does not easily convey enough information for the externalization of sounds; the sound sources do not appear to be out there. In this experiment the turning of the head and the associated change of the relative left and right speaker intensities and delays provide the additional information that gives unambiguous directions for the sounds. A sound is externalized, because a direction is seamlessly associated with it.

This experiment highlights the importance of active exploration in auditory perception. The turning of the head is an explorative act that provides the required additional information about the sound source direction in respect to the head.

6.3. Externalization of Visual Percepts

Natural vision externalizes visual percepts. Eyes project images of seen objects on the retina, but this is not the experienced location of the objects. We do not see the retinal image as a map or a depiction of the outside, instead we have the impression of being able to see and inspect the outside world directly as it apparently is. This impression can be gained via simple explorative experiments; the closing of eyes or covering them with hands will show that the visual world is out there, behind the eyelids or hands. But there is more.

Stereoscopic vision with two eyes creates the impression of depth by allowing the other eye see a little bit of what is behind an object that obscures the line of sight of the other eye. This effect can be amplified by turning the head or moving a little. This will change your vantage point, and you will notice that the objects seem to move a little bit in relation to each other, and the visual scene will no longer be static. You will see which objects are behind others.

Visually seen objects are at different distances and directions from us. As good as the eye is, its resolution is excellent only at a very small area of the retina, fovea, that is directly opposite the lens. The visual angle of the fovea and the accurate vision is only about two degrees, corresponding to four diameters of the seen sun or moon. Therefore, in order to see fine details accurately, we have to scan the view with our gaze. This, however, has a benefit; the gaze direction determines the apparent locations of seen objects, and can be used for the sub-conscious memorization of what is where.

The direct visual impression of the outside world is seamlessly connected to our ability to move; we can reach out and touch seen close by objects effortlessly without any apparent calculations about the required motion trajectories.

6.4. Externalization of Body Sensations

The externalization of sensory information applies also to body and touch sensors. When we touch something with our fingers, the sensation appears to take place at our fingertips, yet the neural activity that produces this impression is located inside the brain. All the sensations that are generated by the various receptors in the skin, have an apparent location, which is the location of the corresponding receptor. Also pain may have an apparent location somewhere in the body, even though the actual feel of pain is created inside the brain. The receptor nerves do not have any intrinsic information about their points of origin and cannot therefore provide enough information for the externalization of their signals. The apparent locations of body sensations and pain outside the brain are related to the internal *body image*, which is created by the combination of information from various sensory modalities during experimentation in early childhood.

The created body image is not as robust as we might believe. The *rubber hand experiment* [Botvinick and Cohen 1998] and other similar experiments seem to verify that the locations of touch sensations are not intrinsically transmitted by the touch nerves, because the perceived location can be quite easily changed. In a typical rubber hand experiment, a rubber hand replica is placed on the table in front of the test subject. The subject's own hand is kept below the table, so that the subject cannot see it. During the test the subject's own hand and the rubber hand are stroked synchronously with paint brushes. The subject sees that the rubber hand is being stroked, and feels the coincident stroking of his own hand. After a short while the subject may begin to feel that the touch sensation originates from the rubber hand; the subject's body image has been distorted. The rubber hand experiment shows that the body image is indeed created by explorative acts and is not an inherent fixed one.

In general, body image information can be achieved by the cooperation of various senses, especially by seeing and touching various parts of the body. A pain sensation without a definite position may be located by touching and feeling the affected body area and noting where the touch sensation and the original pain sensation coincide; "feeling where it hurts". Due to the acquired body image we readily know, for instance, which sensations relate to the conditions in our fingers. Thus, when we cut a finger, the pain will not seem to be in the brain, where it is actually generated. Instead, the pain appears right there in the finger, and will move around with the finger, when we move our hand.

Yet, there may be cases where the body image is not perfect and we must touch and feel, in order to find out where it actually hurts. The externalization of a contact sensor sensation is based on exploration and the resulting association of body part positions with it.

6.5. The Meaning and Interpretation of Percepts

In humans, all receptors produce responses that are perceived as qualia. As explained before, qualia are self-explanatory and indicate the real or virtual presence of the causally connected stimuli. Qualia represent properties, qualities and elementary features, and combinations of these give the impression of different forms and patterns. Initially qualia have no other meanings.

The repeated perception of sensory forms and patterns will allow them to be remembered. In this way unfamiliar forms and patterns turn into familiar objects, but they still lack deeper meanings. A frequently seen pattern is a familiar pattern, but what is its meaning? This depends, see Fig. 6.4.

O 1OO BOOK

Fig. 6.4. What is this round pattern? Is it a circle, a number or a letter? It is all of them and perhaps something more. The context will tell, which meaning is relevant.

Percepts must be interpreted. They must be put into the context of current situation and already known information. This is not a one-to-one process; a sensory pattern may get different meanings in different contexts. Perception processes alone cannot achieve this, therefore additional processes are needed.

Information may be coded into patterns, but the meanings are in the mind of the observer. Percepts can be associated with additional meanings by learning them and having them available from memory. In Fig. 6.4. the basic round pattern may be automatically seen as a geometrical circle, the number zero, the letter o or the symbol for moon, etc., depending on context; what the observer has learned before and what the observer expects. Obviously the basic round pattern has been associated with all of these, but only one association, the one that is relevant to the context, is evoked at each situation. The actual associative evocation process is normally subconscious and effortlessly instantaneous.

Additional associated meanings of percepts cannot be available without memory. Memory is also required for learning, thinking and imagination.

Chapter 7

Memory, Learning, Thinking and Imagination

7.1. Memory Function

Without memory function all actions would be instantaneous stimulus-response actions. Nothing could be learned as nothing could be remembered. The sensed world would not have any perceived order or continuity. Without memory thinking and imagination would be impossible. Nothing could be reasoned, nothing could be planned. There would be no remembered past, no expected future.

Cognitive psychology divides memories into different classes according to their specific function. Usually memories are divided into short-term sensory memories, short-term episodic memories, long-term memories and skill memories.

Sensory memories sustain sensory percepts for a short while. Examples of these are the auditory echoic memory and the visual iconic memory. The echoic memory is able to sustain and repeat a heard sound with its tonal qualities up to couple of seconds, if no interfering sounds are present. The sustaining time for iconic memories is much shorter.

Short-term episodic memories, long-term memories and skill memories are created by the association of things with each other, and can therefore be evoked by cues, parts of the associated things.

Short-term episodic memories sustain the memory of recent happenings for a while. These memories are not as detailed as short-term sensory memories. Short-term episodic memories are necessary for the execution of daily actions. One must remember what one has already done and what should be done next; what things have been taken to where, etc. Short-term episodic memories are volatile and will decay within days.

Some happenings are important and should be retained in memory. The capacity of short-term memories is not sufficient for this, and moreover, old memories might interfere with current memories. For instance, one has to remember accurately where one left one's car at the car park today; previous memories of different locations would only be a distraction. Forgetting outdated information is good, but important matters should be remembered. Therefore, it is useful to copy important memories from short-term memory into long-term memory that has a very long decay time.

In addition to these there are also motor skill memories. Skills must be learned by rehearsal, but the established skill memories are practically permanent.

7.2. Learning and Memory

Learning is the process of acquiring knowledge and skills. Learned knowledge must be memorized and remembered. Without memory function there can only be low level adaptation, not any true learning.

Learning involves the formation of mental connections between things. This can be achieved via association. Simple association connects two things with each other, so that one may later on evoke the other one.

Pavlovian conditioning is an example of simple learning, where two things are associated with each other. The name Pavlovian comes from the Russian psychologist I. P. Pavlov, who in around 1900 pioneered research on simple stimulus-response association in dogs. Pavlov executed a series of experiments, where he rang a bell just before giving food to the dogs. Pavlov noticed that after several repetitions the dogs would begin to salivate when they heard the bell ring, even though no food was present. Pavlov concluded that food and the bell ring had become associated with each other in the dogs' minds. This was later on known as the *Pavlovian conditioning* [Pavlov 1927/1960]. The principle of Pavlovian conditioning is depicted in Fig. 7.1.

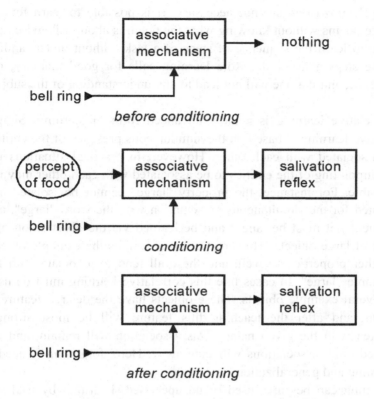

Fig. 7.1. Pavlovian conditioning with an associative mechanism. Bell ring and food are associated with each other in a dog's mind. Before conditioning, a bell ring will not evoke salivation. During conditioning the bell ring is associated with the percept of food. After conditioning, the bell ring alone will evoke salivation as if food were present.

These kinds of associations are present also in humans, and can be easily noticed in daily life. Associative processes may also be the mechanisms behind more complex learning and memory processes.

Rote learning, like the learning of a poem by heart, is memorization by repetition. In the learning of a poem each successive word will become associatively connected with the previous ones, and eventually can be evoked by these. Rote learning can be seen as a sequence of Pavlovian learning steps. Rote learning is not based on the understanding of what is being learned. No meaning is involved, and connections to additional knowledge are not necessary. It is possible to learn foreign language poems without knowing what the poem is about. Likewise, it is possible to learn the sequence of steps of a task without understanding what the steps actually do. Rote learning calls for good memory, not intelligence, and it alone will not lead to any understanding of the subject matter.

Correlative learning is a more advanced form of learning. Simple associative learning is based on the simultaneous presence of the entities to be associated with each other. However, in practical situations the exact pinpointing of the entities to be associated with each other may not be possible. For instance, the property "large" cannot be isolated and pinpointed for the simultaneous presentation with the word "large", and consequently it must be taught and be learned via the observation of a number of large objects. This creates a problem, as the example objects have other properties as well, and these all tend to associate with the given name "large". In cases like this correlative learning must be used with several example objects. These objects have the desired feature in common, and after the teaching this feature will be most strongly associated with the given name. This association will remain, and the unwanted other associations will fade away. Here, forgetting is needed for learning and generalization.

Learning can be supervised or unsupervised. Learning by trial and error is a form of unsupervised learning, where the learner tries different approaches for the production of the desired outcome. This is also associative learning, where certain approaches are associated with bad outcomes and certain other approaches are associated with good outcomes. Learning by trial and error requires the function of judgment. Learning by trial and error is not always safe, and wrong approaches may lead to damage, ridicule and possible injury.

It also possible to learn by observing the actions of others; let the others take the risks and consequences of trial and error. Good approaches should be imitated, and bad ones avoided. Learning by imitation can take place without and with the understanding of the purpose of the imitated steps and actions.

Advanced methods of learning utilize meaning. During learning, connections between the framework of already known and new information are created. The more is already known, the more new connections are possible, and the more cues will be available for the recall of the new information. Learning becomes easier. This kind of learning by connected meanings is faster than rote learning, and is also more useful.

7.3. Memory, Time and Consciousness

Without change there is no time. Without memory there is no sense of change. Memory allows us to remember what was a moment ago and to observe the difference between that and what is now, the change.

The perceived flow of time arises from our organized memories. Time is going forward, because memories accumulate; today we have more memories as yesterday, and this year we have more memories than last year. We also know, which of the lately accumulated memories are more recent than others, because recent memories have more connections to the present situation. Recent memories are usually also more detailed, while old memories tend to fade. The chronological order of recent memories is rather well preserved due to their intact sequential episodic nature. The chronological order of very old memories is not necessarily as easy to reconstruct.

There is past, present and future. They are in our minds. Only our present moment is connected to the real world via sensory perception and physical interaction. The past and future are not. The things of our past are only available by recollection, and the things of the future are only available via imagination; they cannot be sensorily perceived. (The observation of far out stars is the observation of their past, but that is not relevant here).

The brain is able to execute mental time travel (chronesthesia). This is the mind's ability to think not only about the present, but also the past and future. This ability is closely connected to situation awareness and self-consciousness. Patients with hippocampal damage are not able remember past episodes or imagine future. These patients live in constant present moment with a time span of the order of only tens of minutes [see e.g. Suddendorf, Addis and Corballis 2009].

How long is present? Sensory perception is not instantaneous, and the processing and connecting of the collected information takes some time. Therefore the duration of present might be some hundreds of milliseconds. The psychological duration of present is much longer. The experienced present moment stays the same, until something different happens.

Consciousness involves the phenomenal perception of the flow of changes. When nothing changes, time stands still, but so does also consciousness. The brain process that creates consciousness, creates also our perception of time.

7.4. Thinking, Imagination and Reality

Thoughts are a part of our contents of consciousness. Thinking is the process that produces the flow of thoughts. Thoughts are always about something, and accordingly, thinking is a process that operates with meanings. Meanings can be carried in many forms, and therefore this definition does not limit "thinking" to the natural language inner speech. Thinking and thoughts can also be based on sensory experiences, forms and symbols that relate, for example, to the sign language of the deaf, the drawings of an architect or the circuit diagrams of an electronics engineer. Imagination is thinking. Syntactic symbolic information processing without grounded meanings, such as the running of a computer program, is not thinking.

Our contents of consciousness is not limited to thoughts; we are aware also sensory percepts as such, as well as pain and pleasure. According to the above definition, the static awareness of these is not thinking, because no operations with meanings take place. As such, just seeing a rose is not thinking, feeling a headache is not thinking, but the experience of these may lead to various thoughts.

There are different modes of thinking: free running commentary, daydreaming, targeted reasoning and planning. Inner speech is normally present during wakefulness. Free running inner speech self-commentary arises as a response to instantaneous sensory percepts and emotional states. Daydreaming is a form of imagination, where the thoughts revolve around pleasant things and activities that one would like to have or be involved with. Reasoning and planning are systematic thinking that is directed towards the solving of the problems at hand. This form of thinking often requires the concept of time, and may also utilize algorithmic methods, such as formal logic and mathematics.

Imagination is the process of creating mental content that is different from that caused by direct sensory perception or memory recall. The imaginations may be in the form of inner imagery, linguistic stories, actions to be executed, etc. Imagination often involves the mental modification and recombination of what already exists in the sensed environment and in memories, for the possible realization.

The environment and its objects can be utilized in various ways. Gibson [1966] has called these possibilities of utilization affordances. However, the perceived possibility to utilize these affordances is a product of imagination. Stones do not broadcast the possibility to use them as hammers.

Planning uses imagination with a benefit. Imagined plans do not have to be executed at once. The outcomes of the imagined actions may also be imagined, and only those imagined actions that would seem to lead to the desired outcome, should be executed.

The author has proposed some enabling functions for imagination [Haikonen 2005] and these are:

- The evocation of mental representations of imagined objects at different positions; what, where
- The evocation of mental representations of change and motion
- The evocation of mental representations of relation: Relative position, relative size, etc. and relative motion; collide, pass by, take, give, etc.
- Mental modification: Make larger, smaller, rotate, combine, move from one position to another, etc.
- Attention, introspection
- Decision making; the selection between competing imagined scenarios.

Imagination and sensory perception are related. Sensory perception produces limited and incomplete information that has to be augmented by memory and imagination. Sensory information may also be obscured. It is not possible to see behind objects, but it is possible to imagine what is there. In darkness one has to imagine what is out there by limited auditory and visual cues.

It is also a known fact that the stories of eyewitnesses are partly based on imagination; they rationalize and reconstruct memories of situations according to their belief of what happened. This applies to perception in general. We do not inspect our environment carefully, instead we operate with quick cues that evoke imaginations and sub-conscious expectations of what might be.

Our memories are laced with imaginations, and our plans and expectations of future are just products of our imagination. Only our instantaneous percepts are briefly connected to reality, and even these are augmented with imaginations. Social ideologies are products of imagination, and so are our own interpretations of these. One might even say that we live in a world that is more than half imagined. Usually this works fine, and with a little faith apparently even better. But in reality, every now and then our world model fails, and the gap between reality and imaginations becomes sorely apparent.

Chapter 8

Natural Language and Inner Speech

8.1. Natural Language

A natural language is a symbol system that allows the description of real and imagined situations by strings of words and sentences. Natural language is used for communication and verbal thinking in the form of silent "inner speech".

A natural language has vocabulary and syntax. Vocabulary is the collection of the available words, and syntax gives the rules for the indication of those relationships that cannot be conveyed by the basic words alone.

The mastery of a natural language has two parts, namely 1) the ability to understand words and sentences and 2) the ability to produce meaningful sentences. These abilities are related, but somewhat independent. (A learner of a foreign language may notice that it is possible to begin to understand a foreign language before the ability to produce grammatically correct sentences in the same language. Also babies seem to understand speech before they can speak.)

The understanding of natural language text and stories can be tested by asking questions about the subject matter of the text. If these questions cannot be answered, then obviously no understanding has taken place.

According to the generative linguistic tradition, a language is self-sufficient requiring no references to the outside world; all words can be defined by other words and syntax. Vocabulary and syntax can be formalized and incorporated into a computer program, therefore automated language understanding should be easy. The text to be understood could be stored in the computer memory as a data file. When a question about the text is given, the stored text could be scanned and the requested information could be extracted by word pattern matching and statistical or other algorithms. In simple cases this approach will lead to initial success. However, this approach turns out to be insufficient for large arbitrary texts, because pattern matching and statistical computations do not constitute real understanding. This is definitely not the way that is utilized by the brain, either.

The use of a natural language as a self-sufficient autonomous system will not work. Understanding will not be possible without extensive background knowledge that supplies a contextual framework and the necessary general information that is not included in the story itself. As explained earlier, in reality the meaning of words and sentences cannot be ultimately defined by other words, sentences and syntax; this approach would only lead to circular definitions. After all, natural language is used to describe real world situations; therefore the meaning of words and sentences must be grounded to the real world entities. This requirement is especially apparent in syntactically correct, but ambivalent sentences that can only be interpreted correctly with the knowledge of the related situation.

The view of natural languages as symbol systems with real world grounding of meaning has led, for instance, to the author's associative multimodal model of language [Haikonen 2003, 2007]. According to this model, the meanings of words are grounded to the percepts from various sensory modalities, while the syntactic structures arise from the relationships in the described real world situations. The multimodal model of language is also an inner situation model.

Inner situation models of language are based on the fact that natural language sentences are descriptions of a real world situations. These descriptions are able to evoke mental visualizations, that is, inner models of the situation. These mental visualizations should be a familiar experience to anyone reading novels; the reader "sees with the inner eye" the locations and happenings of the story, and also makes memories of these. In doing so the reader actually augments the writer's work by interpreting and "illustrating" it virtually. Without these processes the reader would not able to make any sense of the story. This also allows the paraphrasing and summarizing of the text with one's own words.

The evoked inner situation models are similar to ones that are evoked directly by sensory perception, but may be simpler ones, incorporating only the most salient features and details. Nevertheless they may be internally inspected in the same way as actual sensory percepts.

Inner situation models are also used in real time, in inner speech thinking and overt commenting of current situations. *A real-time perceptual situation model* of the world arises from percepts. Perception produces fleeting percepts that are soon replaced by new ones. However, information about the locations, motions and properties of previously seen, heard or felt objects is retained in short-term memories, and this information with its cross-connections constitutes the model for the current situation. Situation models facilitate *situation awareness*.

A situation model is constantly updated, and it is also associatively linked to long-term memory, background information and emotional evaluation. Situation models change, when situations change. Accordingly, Sommerhof [2000] has called these models "running world models". The connections between the sensed situation, the situation model, its linguistic description and possible action responses are depicted in Fig. 8.1.

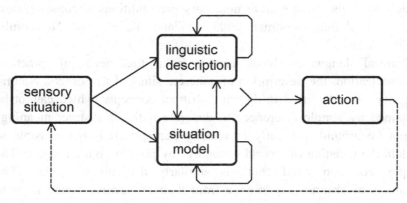

Fig. 8.1. Sensed situation evokes internal interactive linguistic descriptions and imagined situation models. These may evoke action, which may alter the sensed situation. Heard speech evokes directly an internal linguistic description. Beings without language may operate by sole imagination.

External situations can be linguistically described without any situation models, by sensory percepts only, but without models any relations between percepts that are not explicitly observable, cannot be readily included in the description. Without coherent models situations might not be properly understood. Therefore real-time situation models, the "running world models" of Sommerhof, are needed. Sensory perception, linguistic descriptions and inner models interact.

Inner situation models can be evoked directly by linguistic descriptions, sentences and stories. The evoked situation model imagery is then processed as if it were produced by direct perception. Verbal questions about the situation direct attention on the relevant details in the evoked model. Therefore it can be seen that words have also the function of focusing attention. Marchetti [2006] has also proposed and developed ideas about the attention focussing function of words.

Zwaan and Radvansky [1998] see situation models as necessary means for language comprehension, and they oppose the traditional autonomous language view that text comprehension could be achieved by the mental construction and retrieval of the text itself alone, without any mental multisensory models of the described situation. Mental models have also been seen as necessary preconditions to consciousness [e.g. Holland and Goodman 2003, Holland, Knight and Newcombe 2007].

Natural languages have arisen from the need of practical communication, the description of concrete situations to others. Natural languages are also used to describe abstract concepts, which cannot be explained by simple references to concrete entities; abstract meanings cannot be grounded directly to sensory percepts. It is not possible to explain the meaning of "world economy" by pointing out an object. The simple, common word "but" is similarly difficult to define. The meanings of abstract words arise from their associative connections to larger context and their style of use.

There have been some attempts towards the use of grounded language in cognitive machines and robots. The author's robot XCR-1 is one recent example. Earlier on the author has simulated the first version of the Haikonen cognitive neural architecture with a PC and a real video camera. This system was able to use and respond to simple natural language sentences. It was able to learn to recognize a typed sentence as a question and give an answer to it, based on internally evoked imagery of objects [Haikonen 1999]. In a rather similar style, Mavridis and Roy [2006] have devised a situation model for robots by using traditional AI means. This model combines language, perception and action in a way that allows the robot to "imagine" verbally described situations and also to answer questions about perceived and imagined situations.

8.2. Inner Speech and Consciousness

Human mind and consciousness are characterized by a natural language inner speech. Inner speech is the silent self-talk that occurs in our heads when we are conscious. We are aware of our inner speech, which is somewhat similar in appearance to the heard speech. Inner speech is a major component of the contents of human consciousness.

Inner speech is an important form of thinking and is understood as a main cognitive difference between man, animals and traditional machines. In folk psychology inner speech is often equated to thinking, even though also non-verbal modes of thinking exist. In inner speech one may comment one's own thoughts and existence and in this way inner speech is related to self-awareness [Morin and Everett 1990, Steels 2003].

It is known that most of the neural activity inside the brain is sub-conscious. This goes for the inner speech, too. The sentences of inner speech are produced sub-consciously; the words are evoked as a response to the instantaneous mental states, and they just follow each other in a suitable order, effortlessly and without any apparent planning. The end result is the inner speech, virtually heard silent self-talk.

The phenomenon of inner speech seems to make us aware of our own thoughts; if we did not hear our inner speech then how could we know what we think? Indeed, without inner speech our scope of conscious thoughts would be limited to vague imaginations and feelings.

We are conscious, when we have inner speech, because being conscious is about having reportable internal subjective experience, and inner speech is one manifestation of that. This fact might appear as a proof of the idea that language and inner speech are necessary preconditions for consciousness. Certain theories of consciousness build on this and propose that especially self-consciousness would arise from thoughts about thoughts; we need thoughts for the observation that we have thoughts, and only in this way we can become aware of them. These ideas are also known as higher-order-thought (HOT) theories of consciousness [Rosenthal 2004].

Higher-order-thought theories of consciousness lead easily to the conclusion that animals and babies without a language and inner speech cannot be conscious. However, this is not necessarily so. For instance, to be conscious of pain does not require any special mastery of a language. Moreover, there are also other forms of conscious thinking, such as visual and kinesthetic imagination. Thus, a meaningful inner speech may be a strong indication about the possible presence of consciousness, while the lack of inner speech is not a proof of the lack of consciousness.

Conscious robots should have natural language inner speech, because in this way their mental processes would be more similar to ours and we could understand better their train of thought. This, in turn, would allow easier human-robot communication.

Chapter 9

Emotions and Motivation

9.1. Emotions and Feelings

Human conscious experience is colored with emotions. Strong emotional feelings make you feel alive, and this is what consciousness is ultimately supposed to be about; to have the feeling of being alive.

Should robots also feel to be alive and have emotions then? Or, should humans be better off without emotions; we do have abundant examples of bad outcomes also in grand scales caused by emotional rage and attitudes. On the other hand, some emotions would seem to be good for us; therefore, should we try to implement useful emotions in a robot? Or, would it be possible that conscious robots would behave in emotional ways even if those were not explicitly implemented in their design? There are good reasons to consider emotions and emotional feelings also in the context of conscious robots.

No exact definition of emotional states exists, but a general idea may be given by a list of typical emotions:

Anger; Curiosity; Boredom; Desire; Despair; Disappointment; Disgust; Empathy; Envy; Fear; Gratitude; Grief; Guilt; Happiness; Hatred; Hope; Horror; Hostility; Jealousy; Love; Lust; Pity; Pleasure; Pride; Remorse; Sadness; Shame; Shyness; Sorrow; Surprise...

It can be seen that emotions are reactions to everyday situations in life, and in many cases they contain the element of conflict. In deed, emotions are triggered by certain situations, and they involve consciously perceived arousal, subjective feeling and physiological symptoms. The

71

reason for a specific emotion is, however, not always consciously perceived.

The physiological symptoms may include some of the following:

Facial expressions; crying; smiling; laughter; blushing; sweating; paleness; shivering; vocal modulation; changes in heart rate; changes in breathing rate; changes in blood pressure; stomach symptoms...

Many of the physiological symptoms of emotions would seem to be useless or even harmful. Why should a shy person have to sweat or feel sick when delivering a speech to a large audience? Would these kinds of emotional reactions be useful to humanoid robots? A shy clumsy robot fearing to talk might appear funny, but not very useful. Obviously there are some evolutionary reasons for the physiological symptoms of emotions, but these reasons might arise from the unplanned evolution of the central nervous system and its connections to the homeostatic control of the body and not so much from any survival aspect.

Facial expressions of emotions may be useful, because these can be used to communicate emotional states to others so that they can adjust their behavior accordingly.

9.2. The Qualia of Emotions

Emotions are internal states of the mind. It is known that in the brain there are no pain sensors; cortex does not feel pain in the same way as, say, a severed finger. The brain is not a sensory organ, and does not have any sensory receptors. How then could emotional states be perceived?

According to the theory of William James the quality of emotional states arise from the perception of related bodily sensations. Emotional stimuli are perceived by the sensory areas of the cortex, which then transmit the information to the motor cortex that generates the bodily response [LeDoux 1996 p. 80]. In this theory the feeling follows the bodily responses and is related to the perception of these. The theory of James was criticized for the apparent reversing of the cause and consequence; are we afraid, because we run, or are we running, because we are afraid?

In a different way, the Cannon-Bard theory of emotions proposes that the feeling and the bodily responses occur simultaneously. According to Cannon-Bard an emotional stimulus enters thalamus and is forwarded to the cerebral cortex and to the hypothalamus. The hypothalamus generates the bodily response, while the cerebral cortex generates the feeling [LeDoux 1996 p. 84]. However, this theory does not explain, how exactly the feeling is generated.

Nowadays the pathways of emotional stimuli are known rather well. An emotional stimulus enters the sensory thalamus and is forwarded via low and high routes to the amygdala, which generates the emotional responses. The low route is a direct one, while the high route goes via the cortex. The low route facilitates quick responses to emotionally challenging situations, such as danger [LeDoux 1996 p. 164].

Previously it was stated that only percepts can be conscious. Therefore it seems obvious that also emotional states are perceived via their physiological symptoms, which, in turn are perceived via a large variety of body receptors. Thus the qualia of an emotional state would be the combination of the qualia of the corresponding physical conditions. For instance, anger would be perceived as a state with elevated blood pressure, heart rate, tenseness and possible stomach symptoms, etc. This model of emotional perception is presented in Fig. 9.1.

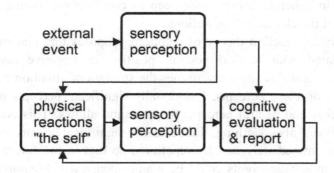

Fig. 9.1. A model of emotional perception. The perception of external events may evoke physical reactions, which are sensed and appear as emotional qualia. Cognitive evaluation may modify these reactions or trigger them alone.

The model of emotional perception of Fig. 9.1 is compatible with the James theory and the Low and High Route model. Sensed external event may excite directly some physical reactions, which are sensed and perceived. The percepts of the external event and the physical reactions are forwarded for cognitive evaluation and the results of this evaluation may modulate the physical reactions. This route allows also the internal evocation of emotional reactions as a result of a thought. The qualia of feelings would be the combined qualia of the physical reactions.

9.3. The System Reactions Theory of Emotions (SRTE)

The author has outlined the so called *system reactions theory of emotions* (SRTE) [Haikonen 2003, 2007], which states that emotions have their roots in elementary sensations. These sensations include good taste and smell, bad taste and smell, pain, pleasure, match, mismatch and novelty. These sensations are characterized by their immediate system reactions. Good taste and smell lead to acceptance, bad taste and smell lead to rejection and withdrawal. Pain leads to withdrawal and possibly aggression. Pleasure leads to sustained action and to the desire of the pleasure-producing activity. Match, mismatch and novelty lead to attention control reactions. More complex emotions arise from the expectations of pleasure (desire, excitement) and pain (fear, excitement) and the outcome of expectations (happiness, disappointment, relief, surprise). In general, emotions are seen as combinations of the system reactions of the elementary sensations.

The system reaction theory of emotions proposes that percepts may be associated with learned neutral, positive or negative *emotional significance*, and this value determines the strength of attention that is to be focused on these percepts. Emotionally significant percepts must be attended first and neutral percepts next, if time allows. This emotional value controls also learning. Emotionally significant events must be memorized immediately, not by repetition, so that in the future proper responses to similar events could be readily generated. Memories have also similar emotional values, the so-called emotional soundtrack that accompanies them. This soundtrack is useful when new actions are planned and their consequences are estimated. The good or bad emotional value of the projected outcome helps to guide the planning towards good results.

SRTE is a practical approach that can be readily implemented in systems that have the necessary elementary sensors and utilize associative processing. Some aspects of SRTE have been experimentally implemented by Kawamura *et al.* [2006] in their ISAC robot. The basic idea of SRTE is also implemented in the author's XCR-1 neural robot.

9.4. The Qualia of Pain and Pleasure

The system reactions approach to qualia combines the external and internal aspects of qualia creation. According to this approach, in each case the special qualities of qualia arise from the system reactions that are triggered by sensory signals. Amodal qualia would arise as a result of amodal sensory responses and other qualia would arise from non-amodal system reactions. Extreme examples of these kinds of qualia would be the feel of pain and pleasure.

Human body is equipped with a large number of pain receptors, which are actually nerve ends that sense cell damage. During a cell damage neural signals are transmitted to the cortex and the feeling of pain is created. The neural signal itself is similar to the other neural signals, which do not appear painful. Where does the hurting feeling come from? The author has proposed [Haikonen 2003] that the hurting feeling arises, when the system experiences its system reactions that are triggered by the pain signal. These system reactions include the disruption of attention. In order to understand this mechanism one has to consider the question: During the perception of a certain feature, a feature signal from the corresponding feature detector is active. The presence of this signal indicates the presence of the corresponding feature, but what else might be perceived about this signal? A simple example illuminates this case.

Imagine listening to your favorite music. Imagine then, that due to some technical problem the music is intermittently switched on and off in a staccato style; after each fraction of second of music there would be an equal period of silence. How would that feel? You would be very irritated, at least. Here you would not only perceive the music, but also the on/off switching of the same; the perception of the music facilitates the perception of the on/off switching, no specific sensors are required. Consider now that the periodic on/off switching of the music would not take place outside your head, instead it would happen in your brain by the switching on and off the corresponding neural auditory signals. It is proposed that you would perceive the situation in a similar irritating way. In addition to the perceived music, you would also perceive the neural signal on/off modulation, which would have an irritating effect. It is proposed that the feeling of pain is related to a similar effect caused by another fast intermittent neural signal that would globally disrupt the attention on the ongoing activities.

Pain is a feeling that demands attention; something must be done to make the pain to stop. In order to get attention, pain would have to suppress other active signals, but not completely, because it would not be safe to make the subject unable to think and do anything at all. Therefore a periodic on/off switching or attenuation of other neural signals would be more useful. The author has proposed [Haikonen 2003] that this is exactly how pain works; it disrupts neural signals in various periodical and chaotic ways, and these disruptions appear as the various feelings of pain. The on/off switching is a system reaction, which is observable within the system without any specific sensors as explained above, and this appears as the quale of pain.

Pleasure demands attention, too. A person wants to continue pleasure-producing action, and is unwilling to shift attention towards anything else. This is different from pain; pain captures attention by disruption, while pleasure is "smooth", it just attenuates other percepts by focusing attention sharply towards the pleasure-producing activity.

9.5. Motivation and Volition

Motivation gives us the reason to do whatever we do in a planned manner, and in this way motivation powers actions beyond simple stimulus-response reactions. Motivation may arise from internal needs and external causes. Internal needs arise from physiological states such as hunger, thirst, fatigue, boredom, etc. An active internal need evokes the want to strive towards the satisfaction of the need.

We want pleasure, and the getting of something that we want (like food, rest and fortunes) gives it to us, or at least we think so during the wanting. Usually we do not want pain and displeasure. In general, the inherent basic and internal motives for action are the avoidance of pain and the seeking of pleasure.

Pain and pleasure are also the basis for the external motivation by reward and punishment. The expectation of a reward pulls the action towards the given goal, while the fear of a punishment has the opposite effect. External motivational factors may be taught and learned.

Reward and punishment relate to pleasure and pain, which are related to emotions, especially to fear (expectation of pain) and desire (want of pleasure). A materialized reward will lead to satisfaction and pleasure, while the not getting of an expected reward will lead to disappointment and displeasure.

In practice, internal and external factors may interact in many ways. An internal need may be acute, but cannot be satisfied until an externally imposed task is completed; a hungry worker cannot leave his post at will without fearing a punishment.

Emotions are important motivational factors in themselves. We do things out of curiosity, out of interest, fear, anger, envy, jealousy, guilt, revenge etc.

A digital computer does not need motivation. A computer executes exactly whatever commands are programmed, whatever conditional branches are provided. The computer itself does not have a freedom of choice.

On the other hand, a true cognitive agent would not be governed by a program. Alternative responses would be possible at any moment, and attention could be directed in numerous ways. No preprogrammed response would exist at any moment, and therefore other means of action planning would be necessary. Motivational factors would have an important role in the shaping of the agent's behavior, and therefore the agent would benefit from emotions. This kind of agent would seem to have some kind of its own will.

9.6. Free Will

Human consciousness is seen to be accompanied by free will. We are apparently free to think what we like, we are free to make our own choices and decisions. Basically this proposition comes from the dualistic view, which maintains that while the material body has to follow the causal laws of physics, the immaterial mind is free from that kind of restriction, and therefore it is free to will whatever it wants. This leads to one argument against conscious machines: Machines can only execute causally controlled deterministic acts. Therefore machines cannot make their own decisions and have free will. Without free will machines cannot be conscious. The obvious weaknesses of this reasoning are the assumptions that 1) free will is a necessary constituent of consciousness and 2) machines cannot execute non-deterministic acts. These assumptions are not necessarily true.

Do we have free will? Free will is the supposed ability to make choices that are free from external influences. In philosophy, an act that is executed by free will is one that does not follow deterministically from the existing conditions, because if it did, then no freedom of choice would have existed in the first place.

We may reject dualism and still maintain that the mind has to have free will, because our everyday experience would seem to prove and demonstrate this. Thus, what kind of experiment could prove that free will exists? This is easy. Tomorrow morning I will prove it by not going to work, even though I would have to. This decision is based on my free will. Quod erat demonstrandum, a philosophical problem is thus solved for good. Or is it? The premises and conclusions of any scientific proof must be scrutinized carefully. Exactly, why did I make this decision? I did it in order to prove the existence of my free will. Thus, my decision has had a cause and obviously only one way of selection; in order to demonstrate my free will, I had to decide not to go to work. If I had decided to go to work as usual, then no point would have been demonstrated, as I had only followed the will of my employer. Alas, my decision was not free, it followed deterministically from the set premises. The existence of free will was not proven.

The author has argued that conscious free will cannot exist [Haikonen 2003 p. 155]. We may buy a certain new car, because its properties suit our purposes and car magazines have recommended it. In fact, we may list our requirements and find out the car that best fits these. This would be our conscious choice. However, this kind of decision making is purely mechanical, and could also be executed by a computer. The point is; in conscious decision making we are aware of the ultimate reasons and steps that lead to the decision. However, these steps are causal, and consequently the process has been a deterministic one, and therefore no free will has been utilized. But, what if we introduced a random factor in the decision making process? Surely this would at least remove determinism. Indeed, we could flip a coin at one step of the decision making chain in order to break determinism, but would the resulting decision be ours then? Of course not. The decision would be determined by the coin and not by us; the decision would not be a product of our free will.

But, if decisions cannot be based on free will, then what are they based on? Does the negation of free will mean that a cognitive robot cannot make its own decisions at all? Does this mean that we would have to provide the robot with decision-making rules for every possible situation? Do humans have decision-making rules for every situation? Not likely.

It can be seen that the decision making process is affected by primary, secondary and external factors. Primary factors arise directly from the situation and include needs, requirements and rational arguments. Secondary factors include emotional state, arousal and alertness, loss of information (forgetting) and own values. External factors include social pressures, the effect of other people, their recommendations and persuasions and any expected reward or punishment. The primary factors are mainly causal and deterministic, and therefore do not contribute to the apparent free will. The external factors reflect the will of others and as such do not contribute to the apparent free will, either. Thus the illusion of free will would arise from the secondary factors that are variable and therefore introduce some kind of unpredictability to the process.

Similar factors would contribute to the decision making process of an autonomous cognitive robot. There would not be pre-programmed rules for each decision, because that would be a practical impossibility. Instead, there should be a general value system that could help the robot to choose to do the right thing.

The apparent free will does have a connection with consciousness. The robot should recognize that the decisions are not something that just pop out from somewhere. Instead, the robot should perceive that a situation calls for selections and decisions, and when these are done, they are made by the robot itself. The robot should see that by doing so and by wanting things, it is able to alter the course of events. Obviously this kind of self-consciousness would create the illusion of an own free will.

It is obvious that no illusions of any kind can arise in program-controlled robots, not even an illusion of free will. Also, it was argued earlier that programmed symbol-processing machines cannot understand meanings. Thinking is manipulation of meanings in the form of "mental imagery". Digital computers run on programs and do only what the program commands make them to do. The brain does not run on programs, therefore a different approach towards cognitive robots should be considered, one that is not based on programs. Artificial neurons and neural networks are this kind of an approach.

Chapter 10

Artificial Neural Networks

10.1. Inspiration from the Biological Neuron

Artificial neurons and neural networks are loosely based on models of biological neurons.

Already at the end of 19th century, Spanish neuroscientist Santiago Ramón y Cajal (1852–1934) was able to show that the brain consists of networks of brain cells, neurons. Based on the work of Cajal and later researchers, a simple model for the biological neuron was developed.

The brain is a biological neural network consisting mainly of neurons, synapses and glia. Neurons and synapses are recognized as the brain's signal processing units, and it has been speculated that artificial cognition could be achieved by reproducing the essential signaling processes of the biological neuron. A simple biological neuron model is depicted in Fig. 10.1.

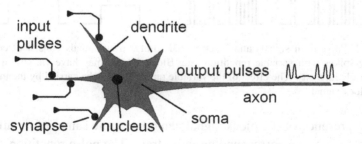

Fig. 10.1. The simple biological neuron model receives input pulse trains via synapses, sums their energies and produces an output pulse train if the sum exceeds a given excitation threshold.

The main parts of the biological neuron are the cell body (soma), dendrites, axon and synapses. The neuron receives excitation in the form of electric pulse trains via the synapses and dendrites. It appears that when the excitation level of the neuron exceeds a threshold, the neuron produces an output pulse train, which is transmitted via the axon.

Simple neuron models see the neuron as a summing unit with a threshold. The neuron receives incoming pulse trains via synapses, multiplies each input signal strength by the synaptic weight of the corresponding synapse and adds the resulting signal strengths together. The neuron will then fire an output pulse train of its own, if the sum of the received pulse train energies exceeds an excitation threshold. This simple neuron model is also known as the *integrate and fire* model, introduced originally by Louis Lapique already around 1907 [Abbott, 1999]. Since then a number of more complicated models have been devised.

When excited, the biological neuron produces trains of equal intensity electric pulses or spikes with a varying repetition rate. Obviously, both the presence of the signal and its pulse repetition rate are carrying information. It is also possible that in some cases also the timing of the individual spikes may convey some information. Artificial neurons do not necessarily have to model and use pulse trains for their operation. Instead of pulse trains also block signals can be used, Fig. 10.2.

Fig. 10.2. Pulse train signals and block signals. Pulse train signals consist of constant intensity spikes with variable repetition rate. Block signals may have variable duration and variable intensity. The effect of the variable pulse rate may be carried by the intensity of the block signal.

The presence of the block signal can convey the same information as the presence of the corresponding pulse train. The pulse repetition rate in a pulse train may convey additional information such as the urgency or importance of the signal. In block signals this information may be encoded in the intensity or voltage value of the signal. The leading edge of a block signal may be used to convey timing-related information. Thus, the information carrying capacities of pulse trains and block signals would seem to be similar. In practical terms, the block signals are easier to implement, and therefore are used in many artificial neurons.

Technically, it should be easy to develop artificial neurons based on the simple model of the biological neuron. But does the simple model actually produce any cognitive functions, and if so, would the produced functions suffice for the artificial production of thoughts?

10.2. Artificial Neurons

First electronic digital computers were developed in the 1940's, and the euphoric feeling at that time was that thinking machines would soon be a reality. But computers run on programs, while the brain apparently does not. Perhaps the non-program imitation of the workings of the biological neuron could be a better approach? Warren McCulloch and Walter Pitts [1943] thought so and produced an early, very simple artificial neuron model. More advanced models, such as the Perceptron were developed later on.

Perceptron is an artificial neuron and neuron algorithm developed by Frank Rosenblatt [1958] at the Cornell Aeronautical Laboratory in 1957, apparently on the basis of the simple McCulloch-Pitts model.

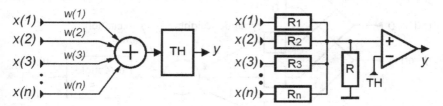

Fig. 10.3. Perceptron neuron principle and one circuit realization, where the synaptic weights are realized with different resistor values.

Perceptron is a summing and threshold unit, somewhat similar to the simple model of the biological neuron. It has a number of inputs and one output, see Fig. 10.3. Each input has a synapse with a given synaptic weight. The intensities of input signals $x(i)$ are multiplied by their corresponding synaptic weights $w(i)$ and are summed together. Perceptron produces output, if the sum exceeds a set threshold TH, else the output is zero. Various threshold functions can be used. Typical examples are the hard threshold, piece-wise linear threshold and sigmoid threshold, see Fig. 10.4. "Soft" thresholds like the sigmoid are used in multilayer neural networks.

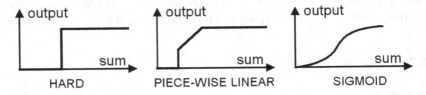

Fig. 10.4. Hard, piece-wise linear and sigmoid thresholds.

Perceptrons can be used for pattern recognition and classification by modifying the synaptic weights. An example of pattern recognition using two perceptrons and binary signals is given in Fig. 10.5. Here the perceptrons recognize "cherry" and "olive" by their roundness and color.

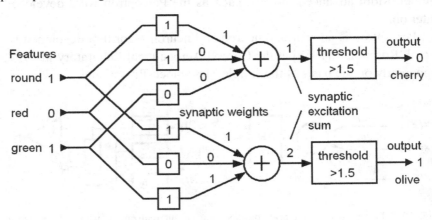

Fig. 10.5. Cherry or olive? Two perceptrons can resolve this.

In the example of Fig. 10.5 detected roundness gives the value 1. Detected green gives the value 1. Red is not detected and has the value 0. These input values are multiplied by the corresponding synaptic weights and are summed together. Thus the cherry-detecting perceptron will get the synaptic excitation sum value 1, and will output zero. The olive-detecting perceptron will get the synaptic excitation sum value 2. This exceeds the output threshold value 1.5, and consequently the output will be 1. An olive has been detected.

The basic perceptron is not very efficient. It is not be able to resolve the so called *Boolean exclusive-or problem*. For instance, the entity to be recognized may have the features A and B, but not both at the same time. There are no such weight values that could discriminate the case, where A and B appear at the same time. However, this problem can be solved by using two perceptron neurons, see Fig. 10.6.

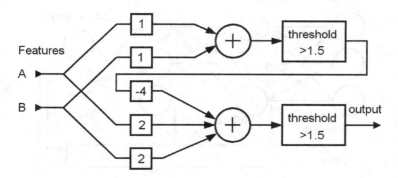

Fig. 10.6. Two perceptron neurons can solve the exclusive-or problem. The upper neuron detects the case A = 1 and B = 1, and prevents any output from the lower neuron.

The circuit of Fig. 10.6. gives output if A = 1 or B = 1, because the lower neuron's synaptic excitation sum of 2 will exceed the output threshold value 1. When A = 1 and B = 1, the synaptic excitation sum will be 2 + 2 - 4 = 0. This remains below the threshold, and the output will remain at zero.

The exclusive-or problem is solved, but at a cost. The circuit setup must be specially planned for this purpose, and the synaptic weights must be externally set, as in the circuit itself there is nothing that could determine these values without external help. Nevertheless, the obvious lesson from this exercise is that improved classification and possibly other advanced functions can be achieved by cascading neurons into multilayer networks.

10.3. Multilayer Neural Networks

A multilayer neural network consists of several successive layers of parallel neurons. Typically there is an input layer, a number of intermediate hidden layers and an output layer. The signals propagate from the input layer towards the output layer, through all neuron layers in between, see Fig. 10.7.

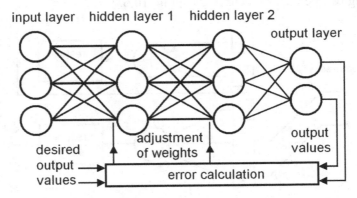

Fig. 10.7. Multilayer neural networks consist of several successive layers of neurons. The signal flow is from the input layer to output layer through the intermediate hidden layers. During training the error between the desired output and actual output is calculated and is used to adjust and tune the synaptic weights.

Multilayer neural networks are used for pattern recognition and classification. Multilayer neural networks learn statistically, and must be trained by using a large number of input examples with their correct results. During training the error between the actual output and the desired output is calculated. Based on this, all synaptic weights are trimmed against each other in the direction that makes the error smaller, again and again. This is done in the hope that the network might eventually be able to produce correct outputs and be able to classify new input patterns on its own. For the adjustment of the weights an algorithm is needed. Most well-known algorithms are the Back Propagation algorithm and the Deep Learning algorithms.

The training of large multilayer neural networks requires usually a large number of training examples. Large networks may also have a very large number of synapses, and consequently the repeated trimming of synaptic weights against each other is a computationally heavy task. The artificial neuron and its synapses as such have inherently nothing that could execute complicated synaptic weight trimming algorithms, and therefore the weight adjustment computations are usually executed by a computer program.

Recurrent neural networks are a variation of multilayer neural networks. In a recurrent neural network the network's output, or a part of it is rerouted back to the network's input via a delay. In this way the network is affected by the current real-time input and the previous output, see Fig. 10.8.

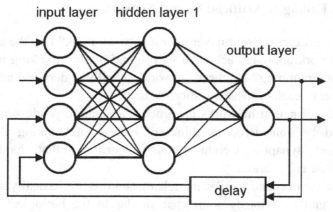

Fig. 10.8. Recurrent neural network. Output is rerouted via a delay back to the input layer.

A fully trained multilayer neural network can be seen as a large look-up table that gives a certain output for each input. A fully trained recurrent neural network can be seen as a set of look-up tables, where the current look-up table is selected by the rerouted output.

Recurrent neural networks can be used as state machines that produce sequences of output patterns. Full computational power can be achieved, if the neurons are made to store and recall intermediate results.

Multilayer neural networks are subject to *the explanatory problem*. Large multilayer networks have many hidden layers, each containing a large number of synapses and neurons. What do the outputs of these neurons represent? How did the network compute its final output? It is not a simple task to find out. There are cases, where a neural network has classified visual information not by the intended features, but by repeating irrelevant background features and even by illumination, and that situation has only afterwards revealed itself, when the network has failed to produce sensible results. Among others, Lieto [2017] has noted that artificial multilayer neural networks do not provide a transparent account of their workings, and moreover, this problem explodes in current Deep Learning techniques. This problem is also known as *the classical problem of opacity* in artificial neural networks.

10.4. The Failing of Artificial Neural Networks

Does a Deep Learning multilayer neural network think? Has the artificial neural network approach achieved its original goal; a thinking machine without programming? Do these networks actually understand anything? The answer to each of these questions is no.

Multilayer neural networks have not been able to produce anything that could be considered as thinking. Instead of thinking they use complicated synaptic weight tuning algorithms that hardly are implemented in the brain.

The Perceptron and similar models of neurons are basically sum and threshold units, obviously somewhat similar to the biological neuron. However, it would seem that the simple function of the Perceptron and the like does not suffice for the production of higher cognition and thinking. Therefore, perhaps the real biological neuron has also another function, one that is not implemented in multilayer networks, but which is necessarily required for the supporting, connection and manipulation of meanings and mental images. If this is the case, then traditional artificial multilayer neural networks are a dead-end alley towards thinking machines, even though they may have been able to produce apparently useful results.

But then, what exactly would be the missing and overlooked additional function of neurons, the function that would enable true thinking? This will be explained next.

Chapter 11

Thinking and Associative Neural Networks

11.1. Thinking as Associative Information Processing

Traditional artificial neural network research has not produced thinking machines despite the efforts to imitate biological neurons and networks. Obviously something has been overlooked, and a certain essential function has remained missing from artificial neuron models. It is clear that the traditional artificial neural networks do not operate in a way that could produce true thinking, but what then would be the correct way? This may be found out by inspecting actual thought processes.

Introspection seems to reveal that our thoughts are not a random stream of separate reflections. Instead of that, each new thought appears to be linked to the previous one by some connection; one thought evokes another via associations. This phenomenon was noted by Aristotle, already some 2500 years ago. Aristotle also tried to find the basic principles of association in human thinking. (A summary of Aristotle's ideas is presented by Anderson [1995].) In modern terms Aristotle's idea could be called *associative information processing*.

Much later on, David Hartley (1705–1757) and David Hume (1711–1776) developed further the philosophical theory of associative information processing or *associationism* as they called it. They proposed that independent ideas in the form of sensations and mental images were associated with each other via contiguity and resemblance. For instance, a present sensory percept may evoke memories of similar occurrences in the past. Simple ideas were considered as kinds of atoms of thoughts. During thinking, the combination of these atoms by the rules of association would give rise to the flow of thoughts.

However, the old theory of associationism has shortcomings that have been pointed out, for instance, by Henri Bergson. In his book *Matière et mémoire* Bergson argued that in a large associative system almost everything would be eventually associated with everything [Bergson 1896/1988]. Therefore one thought would evoke everything else, and no coherent train of thoughts could arise. Bergson stated that contiguity and resemblance cannot provide a sufficient limitation and selection mechanism, and therefore associationism cannot explain thinking.

Despite Bergson's objections, introspection seems to confirm that thinking is at least partly an associative process, and obviously there are additional contributing factors that Bergson failed to see. In fact, contiguity and resemblance are not really the fundamental mechanisms for associative information processing; Pavlovian conditioning and Hebbian learning are. These mechanisms associate two things together via coincidence, and this is the function that the Perceptron and similar artificial neurons do not perform.

11.2. Hebb's Associative Learning

In his book *The Organization of Behavior* Canadian psychologist Donald Hebb proposed a neurological learning mechanism involving two or more neurons. According to Hebb, when a neuron A and a neuron B persistently fire at the same time, a synaptic link is grown between these two neurons. Hebb noted that "cells that fire together, wire together." This proposition is known as the *Hebb's Rule* [Hebb 1949].

Hebb's model explains simple associative learning, where the simultaneous activation of certain neurons leads to the strengthening of the synaptic connections between these. In this way the firings of these neurons will become associated with each other. Later on the firing of some neurons will lead to the firing of the other neurons, too, due to the learned synaptic connection, and in this way the original neural activity pattern is recreated in the group of neurons.

Hebbian learning can be seen as a simple neurological mechanism for Pavlovian conditioning in the context of two or more neurons. Hebbian learning can also be unsupervised and very fast. The principle of Hebbian learning can also be mathematically formulated in different ways. Various Hebbian learning algorithms for the determination of synaptic connection weights in artificial neural networks exist [see e.g. Jain *et al.* 1996]. Hebbian learning algorithms are usually simple and do not necessarily involve the adjustments of synaptic weights against each other.

Hebbian learning is not very suitable for traditional multilayer neural networks, but it can be easily used in single artificial neurons and synapses, if these are designed to be associative in the first place.

11.3. Associative Neurons and Synapses

11.3.1. The Haikonen Associative Neuron

Associative information processing connects signal patterns with each other, and evokes them by each other. This calls for a neural connecting unit, an associative neuron. This neuron would be able to recognize signal patterns and connect these associatively with other signals.

The author has devised an associative neuron (Haikonen associative neuron) along these requirements. This neuron utilizes modified Hebbian learning and implements the function of connection that is missing in Perceptrons and the like. While the basic and actually only function of Perceptron is the learning of a pattern, the Haikonen neuron can do more; it can learn a pattern and its association with a signal, and evoke that signal with the presence of the learned pattern. This associative function also enables the grounding of meaning, if the used signals and signal patterns represent sub-symbolic self-explanatory features. Also, the associative function of this neuron allows the seamless transition from sub-symbolic to symbolic processing, including the use of a natural language with grounded meanings.

The Haikonen associative neuron is also designed to comply with requirements of match/mismatch/novelty detection [Haikonen 1999, 1999b, 2007]. The neuron is self-sufficient, requiring no external computations for the determination of the synaptic weights.

Technically, the Haikonen neuron is designed to associate one signal with a signal pattern. The single signal to be associated is called here the main signal, and the signal pattern is called the associative input pattern. Thus the neuron has one main signal input, one main signal output and at least one associative input, see Fig. 11.1.

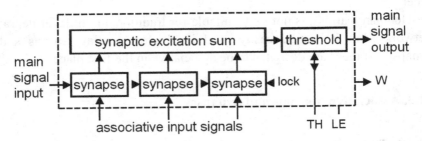

Fig. 11.1. The Haikonen associative neuron has one main signal input, a number of associative inputs and one main signal output. The main signal and the pattern of associative input signals are associated with each other. The neuron associates one associative input signal pattern with one signal. Additional connections may include the threshold value TH, learn enable LE and synaptic weight sum W. LE operates via the lock signal, which prevents the synapses from learning. When W > 1 the neuron has learned a pattern, and lock may be generated internally.

A simplified depiction of the Haikonen neuron is given in Fig. 11.2.

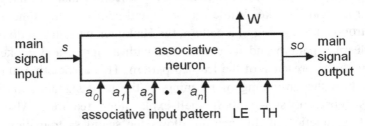

Fig. 11.2. A simplified depiction of the associative neuron. Here s is the main input signal, a_0, a_1, a_2 ...a_n are the associative inputs, and so is the output signal. TH is the threshold value input, LE is learning enable input, W is the synaptic weight sum output.

The Haikonen neuron conserves the meaning of the main signal; the meaning of the main input signal s and the output signal so is the same. This property satisfies the requirements from the grounding of meaning.

Also, as a result there will be no *explanatory problem*, as the meanings of signals can be traced throughout Haikonen neuron networks.

The Haikonen associative neuron is a learning memory unit, where the synapses are the elements that learn and store the associative connection between the main signal and the associative input signal pattern by single or repeated coincidences of these. After learning, the synapses produce an output as their response to associative input signal patterns. These outputs are summed together, and the neuron produces an output if this synaptic excitation sum (*ses*), exceeds a set threshold. In the simplest realization the output of a synapse has two values only, zero or one. The associative synapses can be realized in different ways.

The operation of the neuron and different synapses is described in the following.

11.3.2. A Simple Synapse

The associative neuron needs one synapse for each associative input signal. This synapse can be very simple, consisting of a coincidence detecting multiplier, a one bit memory and a synaptic switch, Fig. 11.3.

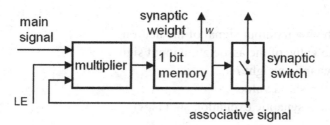

Fig. 11.3. The simple associative synapse consists of the multiplier, the one bit synaptic weight memory and the synaptic switch. The synapse associates the main signal with the associative signal. Association takes place via learning whenever the main signal and the associative signal occur at the same time. This sets the synaptic weight *w* to the logical value 1. This in turn closes the synaptic switch and the associative signal gets a permanent path into the neuron. Learn enable signal LE (0 or 1) may be used to allow learning at specific times. Synaptic weight *w* output is useful in certain applications.

The content value of the one bit memory is called the synaptic weight (*w*) of the synapse, and it controls the synaptic switch. The synaptic weight can have only two values, zero and one. When the synaptic weight value is zero, the synaptic switch is open and the associative signal does not pass the synapse. When the coincidence of the main signal and the associative signal is detected, the synaptic weight value goes to one, the synaptic switch will be permanently closed, and from that point on the associative signal will pass the synapse. The operation of the synapse can also be considered as multiplication; the associative signal is multiplied by the synaptic weight, zero or one.

In this simple synapse already one coincident occurrence of the main signal and the associative signal suffices for the creation of the associative connection between these signals.

In many cases some kind of learning control is necessary, as the learning in given groups of neurons must take place only at suitable occasions determined by the focus of attention. This is facilitated by the learning control signal.

The operation of the neuron with the simple synapses can be formulated as follows:

$$So = 1 \ IF \ \Sigma(a_i * w_i) > TH \quad ELSE \quad So = 0$$

where
  So = the main output signal of the neuron
  a_i = the value of the associative input signal i
  w_i = synaptic weight i
  TH = threshold value
  $\Sigma(a_i * w_i)$ = synaptic excitation sum (*ses*)

It can be seen that the neuron produces output, if the input pattern of a_i values matches the pattern of synaptic weight w_i values. Ideally this should be an IF-AND-ONLY-IF function, but unfortunately with the simple synapse this is not the case. Therefore the simple synapse can only be used in some specific applications. For general applications a more advanced synapse is needed. This kind of a synapse is described in the following.

11.3.3. The Interference-Free Synapse

The Haikonen associative neuron with simple synapses is not perfect as a pattern recognizer, not any better than the Perceptron. Ideally, the neuron should produce output, if the input pattern of a_i values is the same or almost the same as the pattern of synaptic weight w_i values. With the simple synapses this does not always happen. The problem is illuminated in Table 1.

Table 1. An associative input pattern $\{a\}$ is correctly recognized, if it matches the pattern of synaptic weights $\{w\}$. In the simple synapse this is determined by the sum of the products of each associative input signal and its corresponding synaptic weight. If the sum is higher than a set threshold, the associative input pattern is taken as recognized.

w_0	w_1	w_2	w_3	w_4	w_5	
0	1	1	1	0	0	

	a_0	a_1	a_2	a_3	a_4	a_5	$\sum a_i \cdot w_i$
1	0	1	1	1	0	0	3
2	0	1	1	1	1	1	3
3	0	0	1	1	0	0	2
4	1	0	1	1	1	1	2

Table 1 presents a simple example with the pattern of synaptic weights $\{w_0, w_1, w_2, w_3, w_4, w_5\}$ and four different patterns of associative input signals $\{a_0, a_1, a_2, a_3, a_4, a_5\}$. Here the first associative input pattern 011100 matches the synaptic weight pattern 011100, and the synaptic excitation sum value 3 is produced. If the threshold is set to 2.5, then the neuron will produce output, and all is well. Unfortunately, the second associative input pattern 011111 produces also the same sum value, and would be falsely recognized. This happens, as the synaptic weight pattern 011100 is a sub-set of the input pattern 011111. This mode of failure is known as sub-set interference [Haikonen 2007].

It would be useful, if the neuron could respond also to associative input patterns that are not exactly the same as the synaptic weight pattern, but rather similar, close enough (*soft detection*). In principle this could be achieved by the lowering of the threshold value. Unfortunately the simple synapse does not work well here, either. In the Table 1 the example associative input patterns 3 and 4 illuminate the situation. The input pattern 3, 001100 can be taken to be close enough to be accepted, but the input pattern 101111 is definitely not, yet they produce the same sum value.

The output of the simple synapse is determined by the product $a*w$. This product does indicate the match between a and w, but it does not provide any information about mismatch conditions. This shortcoming is behind the problems, but it can be remedied easily by replacing the product a_i*w_i rule with the rule of the Table 2.

Table 2. The synaptic rule for an interference-free synapse for the Haikonen neuron. Zero-zero situation must not lead to a response, as otherwise the neuron might fire also on its own, when the inputs are zero. For each case of penalty, a little bit is subtracted from the synaptic excitation sum (*ses*).

a_i	w_i	result
0	0	no response
0	1	penalty
1	0	penalty
1	1	1

In the interference-free synapse the synaptic on/off switch of the simple synapse is replaced by the rule of the Table 2. This can be realized in several simple ways in actual electronic circuits.

11.3.4. The Exclusive-Or Problem

The Perceptron neuron suffers from the Boolean exclusive-or problem. It is not be able to resolve the situation, where an entity to be recognized has the features A and B, but not both at the same time. Technically, the feature patterns A = 0, B = 1 and A = 1, B = 0 should be recognized as the same entity, while the feature pattern A = 1, B = 1 should not.

Two Haikonen associative neurons with interference-free synapses solve this problem easily, see Fig. 11.4.

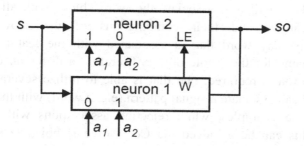

Fig. 11.4. Two Haikonen associative neurons with interference-free synapses connected to solve the exclusive-or problem. Initially the neuron 2 is not enabled.

In Fig. 11.4 the neurons 1 and 2 have common s inputs and common a_1, a_2 inputs. Also the outputs are connected together. The synaptic weight sum output W of the neuron 1 is connected to the learning enable input LE of the neuron 2, so that initially the neuron 2 is not enabled.

The task of this circuitry is to learn to associate the a-input patterns $\{a_1, a_2\} = \{0, 1\}$ and $\{a_1, a_2\} = \{1, 0\}$ with the input signal $s = 1$. This takes place via associative learning. The neuron 1 associates the a-input pattern $\{0, 1\}$ (or $\{1, 0\}$, which ever comes first) with the s-signal, locks, and W goes up enabling the neuron 2. Next, the neuron 2 associates the a-input pattern $\{1, 0\}$ with the s-signal and locks. Thereafter the a-input patters $\{0, 1\}$ and $\{1, 0\}$ will evoke the output signal so, while the a-input pattern $\{1, 1\}$ will not evoke anything.

The use of two Haikonen neurons in the configuration of Fig. 11.4 solves the exclusive-or problem without any ad hoc circuitry. Also the interference-free neurons are able to learn here the required connections by their innate learning mechanisms.

11.3.5. The Correlative Synapse

Simple Hebbian learning is fast, but it is not well-suited for correlative learning. Consider for instance a situation, where the adjective word "blue" is to be defined. One may show a blue object to the associative system and at the same time say the word "blue". An associative system with simple neurons will now associate "blue" with the color blue, but unfortunately it will also associate the word "blue" with all the other detected features and their combinations of the shown object. Consequently, any word that is associated with the feature set, will become a name for the whole object. Therefore a different, correlative learning neuron is required, one that is able to utilize several example objects, and can associate a signal pattern (e.g. a word) with the common feature of the examples, while rejecting associations with the other features. This can be achieved via Correlative Hebbian learning with correlative synapses. A correlative synapse is depicted in Fig. 11.5.

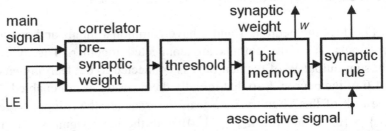

Fig. 11.5. The correlative synapse using Hebbian learning. The pre-synaptic weight goes up and down at each coincidence and non-coincidence of the main signal and the associative signal. If the main signal and the associative signal are correlated, the pre-synaptic weight will eventually exceed the threshold and the actual synaptic weight will be set to "one". Thereafter the correlator circuit has no longer any function.

In Fig. 11.5 the multiplier of the simple synapse of Fig. 11.3 is replaced by a correlator circuit, which accumulates the so-called pre-synaptic weight value. The pre-synaptic weight goes a small step up at each coincidence of the main signal and the associative signal. If the main signal appears alone, the pre-synaptic weight goes a small step down. If a correlation between the main signal and the associative signal exists, the pre-synaptic weight will eventually exceed the preset threshold and the actual synaptic weight w will be set to "one". This value will be stored at the one bit memory. Thereafter the correlation circuit has no longer any function.

In the following the learning of the association of the concept "blue" (for instance, the word "blue") with the feature <blue> is presented as an example. In this simplified case the single main signal of the neuron represents the concept "blue" while a number of associative feature signals represent the various features of perceived objects. Let us assume that three objects are used for the training; these objects are a blue book, a blue flower and a blue pen. All these objects activate the common feature "blue" and also a number of other features that are specific to each individual object. In this example the other features are also represented by one signal for clarity, see Fig. 11.6.

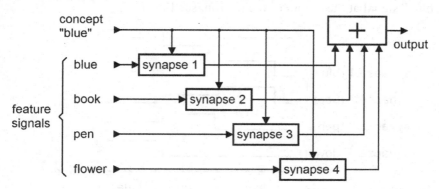

Fig. 11.6. The association of the concept "blue" with the sensory feature <blue> by using the examples of blue pen, blue book, and blue flower.

Figure 11.6 depicts a neuron with four synapses. In this example the main signal is set to represent the concept "blue", while the meanings of the feature signals are determined by the fixed wiring from their feature detectors. The task is to associate the feature <blue> with the concept "blue" and reject all other associations. Thus, only the feature <blue> should evoke the concept "blue" at the output of the neuron, while the features of <book>, <pen> or <flower> should not evoke any output. This will happen, if the synaptic weight of the synapse 1 has the value 1 and the other synaptic weights have the value 0. Correlative Hebbian learning picks up the common feature among a number of objects.

During the training of the neuron of Fig. 11.6 first the blue book is presented simultaneously with the concept "blue". At this moment the pre-synaptic weights for the blue/blue synapse 1 and blue/book synapse 2 step up and others remain at zero. Next, the blue pen is presented. Now the blue/blue synapse 1 steps up again and also blue/pen synapse 3 steps up while others go down. After the last example object, the blue flower, the blue/blue pre-synaptic weight has the highest value and the learning may be completed by using a threshold value of 2.5. The synaptic weight of the synapse 1 will now have the value 1 and the other synaptic weight values will remain at the value 0; the neuron has learned the association between the feature <blue> and the concept "blue" and thereafter the presence of the feature <blue> will evoke the corresponding concept "blue" signal at the output of the neuron, see Fig. 11.7.

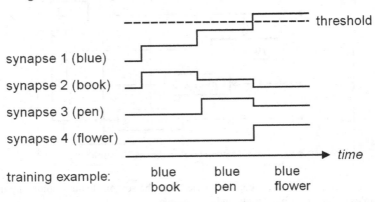

Fig. 11.7. Pre-synaptic weight values during the correlative learning of the association of the concept "blue" with the feature <blue> using the examples of a blue book, a blue pen and a blue flower. After the third example "blue flower", the pre-synaptic weight of the synapse 1, blue/blue, exceeds the threshold and the concept "blue" and the feature <blue> are associated with each other. Other pre-synaptic weights will eventually fade away.

Likewise, the common features of various examples of books could be associated with the concept of "book", while the associations with untypical features, which might be common to totally unrelated objects, would be rejected. In this way Correlative Hebbian learning is useful in the learning of simple meanings of words.

The Haikonen neuron with correlative synapses can also be used as unsupervised pattern learners by setting the main signal input permanently to one. In this case the neuron will learn a repeating pattern or a common sub-pattern on its own power, without the trimming of synaptic weights against each other; complicated synaptic weight computing algorithms are not needed. In this application the learning should be disabled as soon as a number of synaptic weights go to one.

11.4. The Associative Neuron as a Logic Element

Is it possible to compare the associative neuron to a logic element such as the logic AND and OR, and if so, what would be the executed logic function? If the associative neuron executed a standard logic function, would it then be possible to build associative neural networks with the readily available digital microchips? A simple analysis gives the answer.

The output condition for an associative neuron with three associative input signals can be written as follows:

$$so = 1 \text{ IF } a_1 \bullet w_1 + a_2 \bullet w_2 + a_3 \bullet w_3 > threshold, \text{ ELSE } so = 0$$

where

so = output signal, (1 or 0)
a_i = associative input signal, (1 or 0)
w_i = synaptic weight, (1 or 0)
$\bullet$ = synaptic rule operation

Let us assume that all the synaptic weights w_1, w_2, w_3 have the value 1. This would mean that the associative input pattern $\{a_1, a_2, a_3\} = \{1, 1, 1\}$ has been associated with the main signal so. The condition for the output becomes now:

$$so = 1 \text{ IF } a_1 + a_2 + a_3 > threshold, \text{ ELSE } so = 0$$

If the *threshold* is set to 2.5, then *so* will become 1 only if every a_i has the value 1. This corresponds to the logic AND function:

$$so = a_1 \text{ AND } a_2 \text{ AND } a_3$$

If the *threshold* is set to 1.5 then *so* will become 1 only if at least two of the a_i have the value 1. This corresponds to the logic function:

$$so = (a_1 \text{ AND } a_2) \text{ OR } (a_2 \text{ AND } a_3) \text{ OR } (a_1 \text{ AND } a_3)$$

The accepted associative input patterns would be: {1,1,0}, {0,1,1}, {1,0,1}, {1,1,1}. This can be considered as a kind of a "SOFT-AND"; all inputs of the true logic AND are not necessary. Thus an associative input pattern that is an approximation of the original associative input pattern, is also able to evoke the associated main signal. This is a useful property in many practical applications, as it categorizes associative inputs and automatically applies learned associations to similar novel situations. SOFT-AND allows also some random variation in the associative input pattern.

If the *threshold* is set to 0.5 then *so* will become 1 if at least one a_i has the value 1. This corresponds to the logic OR function:

$$so = a_1 \text{ OR } a_2 \text{ OR } a_3$$

OR function is also useful in many practical applications.

Thus, it can be seen that by varying the *threshold* value the associative neuron can be made to execute different logic functions according to the requirements of each application. For the sake of clarity and simplicity an associative pattern with three signals was used in the previous examples, but it should be obvious that these principles apply to associative patterns of any length.

Thus, it can be determined that in those applications, where the true AND or true OR is to be executed, standard logic circuits could be used. There are no standard microchips for the SOFT-AND function.

However, the associative neuron has also other functions in addition to the logical ones. The associative neuron is able to learn and associate signals with each other. This is implemented by simple Hebbian synapses, interference-free synapses or the more complex correlative Hebbian synapses. Standard microchips that execute both the associative learning function and the logic function are not currently available.

Practical associative systems require associative neuron groups that are able to associate signal patterns with one signal, one signal with a signal pattern and signal patterns with signal patterns. These neuron groups are described in the following.

11.5. Associative Neuron Groups

11.5.1. The Association of a Pattern with One Signal

One associative neuron is able to learn the association between one associative input signal pattern with one signal. This can be formulated as follows:

$$\{a_0, a_1, a_2, \dots a_n\} \rightarrow so$$

where the pattern $\{a\}$ is the input associative signal pattern and *so* is the associated signal. After the learned association the $\{a\}$ will evoke the associated signal *so*.

One neuron is able to detect one pattern or patterns that are quite similar. Therefore a neuron group is required, if a number of different $\{a\}$ patterns are to be associated with their corresponding *so* signals, see Fig. 11.8.

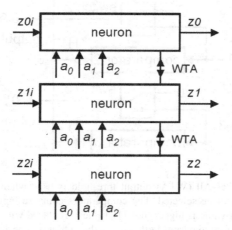

Fig. 11.8. An example of an associative neuron group with three neurons and common associative input lines. An input signal pattern, $\{a\}$, is associated with one output signal.

The situation is more complex, if partial similarity between the $\{a\}$ and $\{w\}$ (synaptic weight) patterns are allowed. In that case a certain input pattern $\{a\}$ may evoke output from more than one neurons, while only one of these outputs would be the best response. For these cases a mechanism must be devised that identifies the best response and subdues the other responses. This can be done by the so called Winner-Takes-All (WTA) threshold control.

11.5.2. The Winner-Takes-All Threshold Circuit

The Winner-Takes-All (WTA) output threshold circuit is used, when several neurons are used as a group, and only the neuron that has the highest synaptic excitation sum is allowed to send its output signal. Figure 11.9 presents the principle and an example of the Winner-Takes-All output threshold circuit.

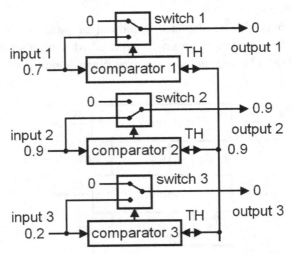

Fig. 11.9. Winner-Takes-All (WTA) output threshold is used when one output from a number of neurons is to be selected. The comparators compare input values to the TH value, and if the input value is higher, the TH value is replaced with the input value and the input signal is allowed to pass. At other times the output will be zero.

In Fig. 11.9 the inputs 1, 2, 3 receive the synaptic excitation sum values from the corresponding neurons 1, 2 and 3. The input 1 has the value 0,7, the input 2 has the value 0.9 and the input 3 has the value 0.2.

These values are compared by the comparators 1, 2 and 3 to the threshold value (TH), which is common to the all neurons of the group.

The threshold value connection line TH is bi-directional, that is, each comparator may receive or transmit a threshold value TH. Normally the TH value is received, but if the synaptic excitation sum of a given neuron is higher than the received TH value, then the comparator transmits this higher value to the TH-line. Due to this arrangement, only one controlling TH-line is required regardless of the number of interconnected neurons.

In Fig. 11.9 the input 2 has the highest value 0.9, which is passed to the threshold line as the valid threshold value. Inputs 1 and 3 are below this value, and are blocked. Only input signal 2 is allowed to pass, while the other outputs remain at zero. The outputs of the WTA are mutually exclusive, "winner takes all".

The WTA circuit of Fig. 11.9 operates in linear mode, where the output signal preserves the amplitude of the selected input signal. Some applications use non-linear mode of operation, where full output value (logical 1) is used as the selected output instead of the actual input value. This can be achieved easily by using the comparator outputs as the actual outputs.

11.5.3. The Association of One Signal with a Pattern

In certain cases a signal pattern must be associated with one signal. After learning the signal shall be able to evoke the associated signal pattern. This can be formulated as follows:

$$z \longrightarrow \{s_0, s_1, s_2, \ldots s_n\}$$

where z is the input associative signal and $\{s\}$ is the associated pattern. After the learned association, the signal z will evoke the associated pattern $\{s\}$.

The above association of one signal with a pattern can be done by a group of neurons, see Fig. 11.10.

Figure 11.10 depicts an example, where a group of three neurons is used to associate signal patterns $\{s\}$ with corresponding associative single signals (*zo*, *z1*, *z2*). These *z* signals are mutually exclusive; only one of them may be active at a time. Each active associative signal can evoke only its own *so* output signals, while the other associative signals are zero and are unable to evoke anything. Therefore this kind of a neuron group has no false output interference problems.

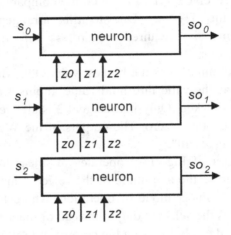

Fig. 11.10. An example of a neuron group for the association of one signal at a time with a pattern. In each neuron the synaptic sum value for each *z0, z1, z2* can only be zero or 1, therefore a simple fixed value output threshold circuit can be used. This kind of neuron group is a neural one-to-n-line decoder with learning ability.

11.5.4. The Association of Patterns with Patterns

The associative neuron group of Fig. 11.8 is able to associate an associative input pattern with one main signal without interference, but it cannot be expanded to handle pattern with pattern association without limitations. This is a shortcoming, because usually full signal patterns must be associated with each other, so that a signal pattern can evoke another, as follows:

$$\{a_0, a_1, a_2, \dots a_n\} \rightarrow \{s_0, s_1, s_2, \dots s_n\}$$

The shortcoming of the simple neuron group can be remedied by the arrangement of Fig. 11.11.

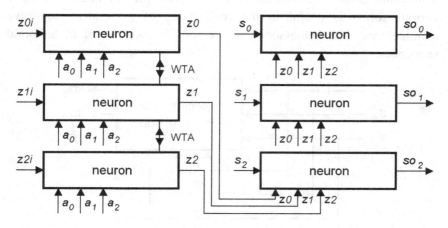

Fig. 11.11. The association of patterns with patterns. This arrangement associates the input signal pattern $\{a\}$ with the output signal pattern $\{so\}$ with two steps in order to avoid interference. First, the input signal pattern $\{a\}$ evokes a single signal z_i, and next, this single signal evokes the output pattern $\{so\}$.

The arrangement of Fig. 11.11 associates an input signal pattern, pattern $\{a\}$, with a signal pattern, pattern $\{so\}$. This is done with two steps. First, a single signal zi is associated with the $\{a\}$ pattern. Next this signal is associated with the output pattern $\{so\}$. Thereafter the $\{a\}$ pattern will be able to evoke the output pattern $\{so\}$. No spurious outputs occur as the both steps are interference-free.

11.5.5. Autoassociative Memory

The processes where two different patterns are associated with each other is called *heteroassociation*. In *autoassociation* parts of a pattern are associated with the whole pattern.

The associative neuron group can be wired to produce autoassociation, too, as an autoassociative memory. In this application the main signal pattern is used also as the associative input pattern, see Fig. 11.12. The main signal pattern S will thus be associated with itself.

The use of *soft detection* allows the evocation of the whole input pattern *S* as the output pattern *So*, when a small arbitrary subset, even with some distortions, is introduced as the input. The principle of autoassociation in various forms is useful in the realizations of associative cognitive architectures. The autoassociative memory is also useful in temporal sequences.

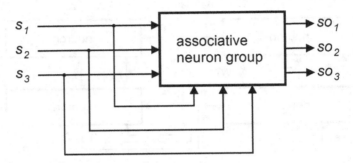

Fig. 11.12. An associative neuron group wired as an autoassociative memory. Here the associative neuron group associates the input pattern with itself. A part of a learned input pattern will evoke the full original pattern, if soft detection is used. The associative neuron group must be configured to be interference free.

11.5.6. Temporal Sequences

Associative neuron groups can only associate simultaneously present signals and signal patterns with each other. However, perception and thinking are temporal processes. Thoughts and percepts are successions of patterns; events follow each other as a sequence, and must be learned and remembered as such. Not only the temporal order of the events in a sequence is important, many times also the duration of the individual events count and must be remembered and reproduced. For instance, a spoken word is a sequence of phonemes of different durations. Moreover, these durations are relative. A word can be spoken slowly or fast, yet it must be recognized as the same word. Therefore, the processing of temporally sequential events calls for the association of non-simultaneous signals and signal patterns with each other, and also means for the recording and replay of the absolute and relative durations of the individual events.

As non-simultaneous signals and patterns cannot be associated with each other, they must be transformed into temporally coexistent parallel forms, where the individual signal patterns are available at the same time. This can be done by short-term memories or delay lines.

Figure 11.13 gives an example of a delay line based circuit that associates a short sequence with a cue pattern.

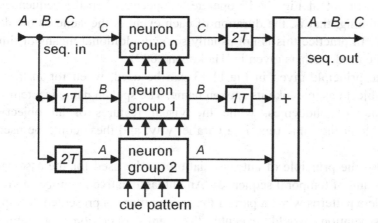

Fig. 11.13. An example of an arrangement for the association and associative evocation of a sequence with a cue pattern. During learning the cue pattern is associated with three successive *A, B, C* patterns of the input sequence. During replay the cue pattern evokes these successive patterns at the same time. However, they are output serially due to the *1T* and *2T* delay lines.

In Fig. 11.13 the input sequence *A-B-C* consists of individual *A, B* and *C* signal patterns, groups of distributed feature signals that appear serially. The cue pattern is a signal pattern that is to be associated with the whole input sequence, so that later on it can be used to evoke the *A-B-C* sequence as the output. Three neuron groups are used in this circuit. The neuron group 0 associates the *C*-pattern of the input sequence with the cue, the neuron group 1 associates the *B*-pattern of the input sequence with the cue and the neuron group 2 associates the *A*-pattern of the input sequence with the cue. The units *1T* and *2T* are delay lines that delay the input sequence so that the *A, B* and *C* patterns appear simultaneously at the inputs of the neuron groups. The actual association takes place, when the last pattern, the pattern C, of the sequence is received.

During the replay of the sequence the cue pattern evokes these successive patterns at the same time. However, they are output serially due to the $1T$ and $2T$ delay lines. the pattern A is output first, then the pattern B appears after the $1T$ delay and finally the pattern C appear after the $2T$ delay. The summing unit passes only the most recent pattern so that the A, B and C patterns will not overlap.

The circuit of Fig. 11.13 operates properly, when the sequence has fixed timing, that is, the duration of each event is the same as the delay time T. In practice this is not usually so, and additional means of timing must be used, such as given in [Haikonen 2007].

The principle given in Fig.11.13 can be used, when for instance, a seen object has to evoke its spoken name. The spoken name is a temporal sequence of phonemes, while the visual features of an object are available at the same time in a parallel way, and these could be used as the cue.

Also the principle of autoassociation can be used for the associative processing of temporal sequences. An autoassociative memory can recall complete patterns when a part of the same pattern is presented. Temporal autoassociation would execute the same operation for temporal sequences; an individual pattern of the sequence would evoke the next one that follows. Autoassociative temporal memory can be used for rote learning and skill learning, where the skill consists of sequences of action. It can also be used for the prediction of what will come next.

The associative synapse, neuron and associative neuron groups are building blocks for the artificial associative realization of the previously described cognitive functions, beginning with perception. In the following it is outlined how this can be done.

Chapter 12

Towards Artificial Cognitive Perception

12.1. Requirements for Cognitive Perception in Robots

Without sensory perception processes a robot cannot acquire information about and become aware of the external world, the robot itself and its own mental content. Therefore perception is a necessary function for any cognitive robot that is intended to be conscious to any extent.

Perception systems for artificial cognitive and potentially conscious robots should be able to produce self-explanatory forms of information for the grounding of the meanings of symbols. This leads to the requirement of direct, non-symbolic, transparent representation of information, which also preserves the amodal features of the sensed phenomena. The perception system must also allow the introspection of mental content.

The perception system must also comply with the chosen way of information processing. Here neural associative information processing with distributed representations is chosen.

A cognitive robot may have a variety of sensory systems for the detection of various physical phenomena, such as light, sound, touch, chemicals (smell and taste), temperature, etc. As discussed before, these physical phenomena as such are not suitable forms for information processing in the cognitive machinery. One cannot input photons, vibrations of air pressure or chemicals directly into a computer or an artificial neural network. Therefore the first step for perception is transduction. This transduction is performed by a sensor, which transforms the sensed phenomenon into the common internal form that is used inside the cognitive machinery. In robots electric signals would be used as the common internal form. The perception process for artificial cognition is presented in Fig. 12.1.

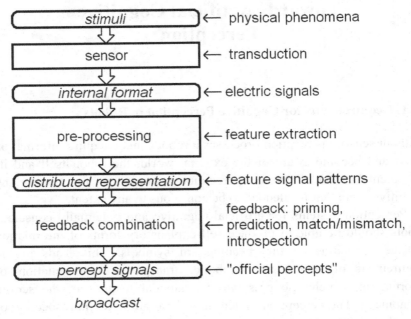

Fig. 12.1. Perception process for artificial cognition.

The next requirements arise from the style of the representation of information. Associative information processing with distributed signal representations calls for pre-processes that extract feature signals from the sensory information. Feature signals are self-explanatory; the presence of a feature signal stands for the presence of the corresponding physical feature.

Sensory perception is not a simple stimulus-response process, it is also affected by the internal state of the cognitive system. Therefore raw percepts must be subjected to the effects of the internal factors such as priming, attention, prediction and the like. These factors would relate to the selection of context related information, augmentation of information by experience, searching for certain information, effects of expectation and so on. For his purpose the sensory process must be able to receive feedback from the inner processes. At the feedback combination point the sensory feature signal patterns can be compared with the feedback signal patterns; this will satisfy the requirement of the detection of match/mismatch/novelty conditions.

The requirements for sensory perception and the introspection of internal mental states might seem to be completely different, perhaps calling for separate circuits and processes, but that is not the case. The feedback can also make mental content available for the sensory perception process by translating inner signal patterns into sensory signal patterns. In this way the introspected mental entities will appear as sensory percepts, but not necessarily with the full and vivid details of actually perceived objects.

The distributed sensory signal patterns that are modified by the feedback, are the "official" output of the perception process. Here these signal patterns are called percepts. These percepts are forwarded or broadcast to the rest of the cognitive machinery.

The assessment of the perception process in Fig. 12.1 leads to the sub-system architecture of a perception/response feedback loop. This sub-system architecture realizes the above requirements for cognitive perception and also provides additional functions, such as short-term memory and the transition from sub-symbolic to symbolic processing by facilitating the association of additional meanings with the percept signal patterns. The perception/response feedback loop is described in more detail in the following.

12.2. The Perception/Response Feedback Loop

12.2.1. The Basic Principle

The general requirements of perception can be satisfied by a feedback (also known as re-entry) circuit arrangement, which is here called the perception/response feedback loop. This circuitry receives information from a sensor and also from the output of the internal neural processes in the form of feedback signals. The internal percept arises from the combined effect of the sensory and feedback signals. The perception/response feedback loop can be constructed by using the earlier described associative neurons and neuron groups. The principle of the perception/response feedback loop is depicted in Fig. 12.2.

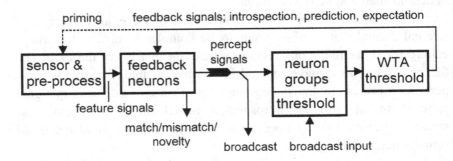

Fig. 12.2. A simple perception/response feedback loop using associative neurons and neuron groups. Pre-processed feature signals are forwarded to the feedback neuron group, which also receives signals from the output of the internal neuron groups. The output signals from the feedback neurons are called percept signals and are broadcast to other units. Here each line depicts a number of parallel signal lines.

Figure 12.2 depicts the general structure and principle of the perception/response feedback loop. Information acquisition is performed by a sensor, which forwards its signals to a pre-processing unit for elementary feature extraction. Each elementary feature is carried by one or more signals. The presence of a feature signal signifies the presence of the corresponding sensory feature. The intensity of the signal may be used to indicate its importance. Importance may also be indicated by using a variable number of signals for the same feature.

The feature signals from the sensory pre-process are forwarded to the feedback neurons. The feedback neurons receive information also from the internal processes of the system. This information is in the form of feedback signals, and may act as priming, expectation, prediction or introspection. Match/mismatch and novelty signals are generated on the basis of the relationship between the feedback and the sensory information. The output signals from the feedback neurons are called the percept signals.

A complete cognitive system utilizes many perception/response feedback loops, which are associatively cross-connected to each other.

The percept signals are broadcast to the other perception/response feedback loops, and are also forwarded to the internal neuron groups. These internal neuron groups receive broadcast inputs from the other perception/response feedback loops. The input thresholds of the neuron groups determine, which broadcasts, if any, are accepted at each moment. The neuron groups are able to learn and remember associative connections between the percept signals and the received broadcast signals. The neuron groups are also able to learn and remember temporal sequences.

The neuron groups may generate a large number of signals as candidates for the output, but only one group of signals can be accepted at each moment. The Winner-Takes-All threshold (WTA) at the output executes the selection, and passes only the strongest candidate signals at each moment. These output signals will be returned to the feedback neurons as the feedback signals.

The feedback loop is able to re-circulate and sustain percept signals for a while after the actual sensory stimulus has disappeared. Therefore the percept points are also perceptual short-term working memory locations.

12.2.2. The Realization of Priming

The so-called *priming effect* occurs, when previously learned percept patterns affect the perception process. The priming effect can occur in different sensory modalities.

The priming effect can be positive or negative. Positive priming helps to recognize patterns during noisy conditions, while negative priming may prevent recognition. An everyday example of negative priming is a situation where a person is searching for a certain object and fails to see it, because the priming has provided expectations of a different-looking object.

Primed perception in the perception/response feedback loop is depicted in Fig. 12.3.

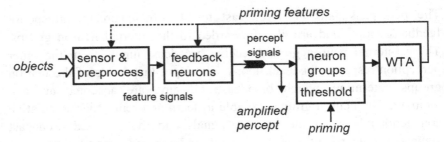

Fig. 12.3. Primed perception. Percept signals that match the priming signals are amplified.

In the perception/response feedback loop the priming effect is facilitated by the feedback signals. Priming is mainly effected at the feedback neurons, but may be extended to the sensors, too.

During positive priming the priming features more or less match the sensory feature signals, and the combined signals will be stronger than the non-primed signals. Therefore the primed signals will win other signals and will be forwarded as the percept signals. During negative priming the priming does not match the actually desired sensory features and no signal amplification takes place for these features.

The priming signals may originate from the internal neuron groups as a result of autoassociation; a part of the sensory percept may evoke autoassociative expectations about the whole sensory pattern. Priming signals may also be evoked internally by signal patterns from other perception/response feedback loops.

12.2.3. Sensory Attention in the Perception/Response Feedback Loop

Sensory pre-processing may produce a large number of feature signals, of which only a limited set can be at the focus of attention at each moment. Sensory attention determines, which feature signals are forwarded to the system, where they may evoke further associations. In the perception/response feedback loop sensory attention may be controlled by external and internal factors, see Fig. 12.4.

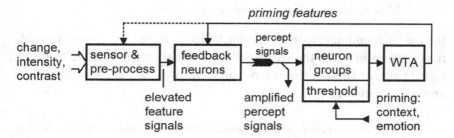

Fig. 12.4. Attention in the perception/response feedback loop may be guided externally by change, high intensity and contrast in the sensed entity. Internally attention may be controlled by priming. Attended percept signals are stronger and may therefore win other percepts elsewhere in the system.

External sensory attention guiding factors include novelty and sudden changes in the sensed modality, higher intensity and contrast. In the auditory modality attention would be captured by sudden unexpected sounds, changes in ongoing sounds, loud sounds and sounds that are very different from other sounds. In the visual modality attention would be captured by bright objects, objects with high visual contrast and by changing and moving objects. Sensory feature pre-processors should be designed in such a way that these properties would cause elevated signal levels for the related feature signals. Thereafter these signals would then appear as amplified percept signals.

Internal sensory attention guiding factors include context and emotional significance. These factors operate associatively via the neuron groups inside the perception/response feedback loop. Context and emotional significance may evoke certain feature signals that are fed back to the feedback neurons as priming features. This leads to the amplification of the primed feature signals and the selection of these by the output threshold circuit. The primed feature signals would then appear as amplified percept signals.

Percept signals are broadcast to the rest of the system. There they will have to compete against other broadcasts at various threshold circuits. The amplified percept signals will win the other signals more likely and therefore may become the focus of global processing and in this way may win the global attention of the system.

12.2.4. Match, Mismatch and Novelty Detection

In the perception/response feedback loop there are three possible relationships between the sensory feature signal pattern and the feedback signal pattern. These relationships can be resolved at the feedback neuron group and they are:

- The *match condition*: the sensory feature signal pattern is similar or almost similar to the feedback signal pattern.
- The *mismatch condition*: the sensory feature signal pattern is not similar or almost similar to the feedback signal pattern.
- The *novelty condition*: Sensory feature signal pattern is present, no feedback signal pattern is present. No expectation or prediction exists.

In simple systems the match, mismatch and novelty conditions may be indicated by binary signals that are generated at the feedback neuron groups. During match condition the match signal is set to one, other signals are zero. During mismatch condition the mismatch signal is set to one and other signals remain at zero. During novelty condition no feedback signal pattern is present, while a sensory feature signal pattern is present. In this case the novelty signal is set to one and the other signals remain at zero. When no sensory signals and feedback signals are present, the match, mismatch and novelty signals remain at zero.

Complete match between the sensory feature signal pattern and the feedback pattern may be rare, therefore the decision between the match and mismatch state must be based on the relative number of individual signal matches and mismatches.

It should be noted that the signals in the sensory feature signal pattern and the feedback signal pattern represent elementary features and therefore can be compared with each other. If, for instance, these patterns were to represent pixel maps, then the comparison would not be feasible; match condition would rarely occur even in those cases, where the images represented by the pixel maps were quite similar.

Match, mismatch and novelty detection is necessary for various cognitive tasks, such as prediction, attention control, search operation and answering "yes" or "no" to questions like "is this xx?".

Match, mismatch and novelty conditions are also related to emotions; match is related to pleasure, mismatch is related to disappointment, novelty is related to surprise, and so on.

12.2.5. The Realization of Visual Searching

During a visual search a given object is looked for. The visual modality perception/response feedback loop facilitates the search operation easily.

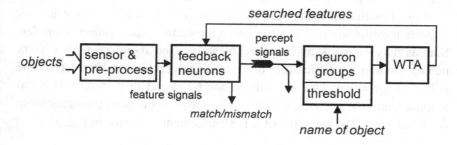

Fig. 12.5. Search operation. The name of the searched object evokes the visual features of the object. These are fed back to the feedback neuron group. When the searched object is found, the feedback matches the perceived object and a match-signal is generated.

In Fig. 12.5 an "inner image" of the searched object is evoked in the visual modality perception/response feedback loop, for instance, by the name of the object. This "inner image" is not a pixel map image of the searched object, instead it consists of a set of more or less invariant features that might match the searched object generally. These searched features are forwarded to the feedback neurons, where they may or may not match the sensory features of the seen objects at the given instant.

When the searched object is found, the sensory features will more or less match the feedback features, and a match-signal will be generated. This signal indicates that the search has been successful, and can be terminated. Otherwise a mismatch-signal will be generated, and the search will continue.

It should be noted that in associative processing also more general visual search methods exist. For instance, one may be in a foreign city looking for a nice restaurant for lunch. In this case one does not have any pre-existing inner image of the searched restaurant. Instead, new visual percepts may or may not evoke ideas of a restaurant and some locations may remind of a restaurant. This mode of operation calls for the cooperation of several modules.

12.2.6. Predictive Perception

Sensory perception is not always a simple process of detecting what is out there. In the visual world objects may be partially covered and shadowed or the illumination may be bad, all this leading to the loss of important information. The situation in the auditory perception is similar. Sounds may be weak and masked by noise, and therefore their correct perception may be difficult. In these cases perception can be aided by the help of experience. This experience can be stored in parallel and serial autoassociative memories, which can be added to the perception/response feedback loop. The principle of this arrangement is shown in Fig. 12.6.

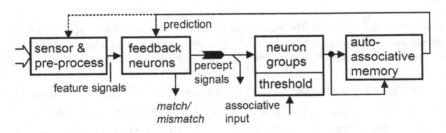

Fig. 12.6. The addition of an autoassociative memory allows prediction in the perception/response feedback loop. Match/mismatch signals indicate the success of prediction.

In Fig. 12.6 the perception/response feedback loop has an additional autoassociative memory. The input of this memory consists of the percept signals or the signals that are evoked at the neuron groups in front of the associative memory. The output of the associative memory consists of the evoked signal patterns, or in the default case when nothing is evoked, the input signal patterns of the memory. This output is returned via the feedback loop back to the feedback neurons. At the feedback neurons the predicted signals can be compared to the actual sensed feature signals. The result of this comparison can be indicated by match/mismatch signals.

In the visual modality a parallel autoassociative memory would enable the prediction of the unseen parts of a known object. A temporally serial autoassociative memory would allow the prediction of motion, the next position of a moving object, when the object moves in ways that are already known to the system.

In the auditory modality a serial autoassociative memory would be used. This would allow the prediction of the continuation of the sound. In linguistic systems the next word could be predicted, and as this prediction is returned via the feedback loop to the percept point, this would again evoke the next word and so on; the whole string of words would be recalled. This has been demonstrated in [Haikonen 1999].

Autoassociative memories allow the perception/response feedback loop to execute simple experience-based sensory prediction, as described above. *Cognitive prediction* is more complicated and involves the coalition of the experience of several modules. Examples of cognitive prediction would be, for instance, the prediction of weather: "Black clouds are gathering in the sky; what may happen next?" or the prediction of future: "What will happen, if I spend more money than I earn?".

12.2.7. Introspection of Imaginations

Imagination is the mental perception and manipulation of actions and entities that are not physically present. Imagination utilizes similar representations that result from direct sensory perception, but in this case these representations arise from and are modified by internal causes. However, internal neural activity is not observable as such, and is in this way "sub-conscious", because, as stated afore, only percepts can be consciously observed. Therefore, the internal products of imagination must be translated into virtual visual percepts that can be handled by the perception process.

The visual perception/response feedback loop module provides a mechanism for the translation of inner representations into virtual visual percepts, and in this way it facilitates introspection of mental imagery. This is done by the feedback that conveys internal neural activity patterns back to the perception process, see Fig. 12.7.

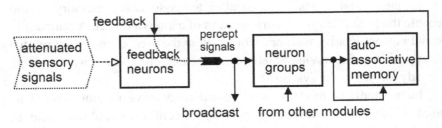

Fig. 12.7. The perception/response feedback loop facilitates the conscious introspection of imagined entities as virtual percepts. If sensory signals are attenuated, the percepts represent the feedback signal patterns in terms of sensory percepts and qualia. In practice, a number of perception/response feedback loops would work together by broadcasting and receiving broadcasts from the other loops.

Visual sensory signals and feedback signals combine at the feedback neurons. When sensory signals are attenuated, the feedback signals will dominate, and will become the official percepts. Visual percept signals have the appearance of patterns and gestalts with qualities, and these will also be the appearance of the virtually perceived imaginations. These imaginary percepts would be emotionally evaluated and broadcast to other modules just like normal percepts. The other modules, in turn, would generate responses and broadcast those, also back to the visual module. In this way a flow of imagery would follow.

The reuse of the perception/response feedback loop for imagination allows also the coupling of imagination to actual perception. Imagined objects and actions can be "overlaid" on top of real perceived environment. This is a very useful property; in fact, without this connection the usefulness of imagination would be rather limited. For instance, it is possible to see an object at a certain location and imagine an action that takes this object to another location. This is a simple imagined plan that is directly coupled to the perceived environment.

12.2.8. Introspection of Inner Speech

Inner speech arises from sensory percepts and internal associations. It also sustains itself; one thought leads to another. All these linkages take place in the inner associative neuron groups, and their products are not directly available for perception. Therefore the inner speech has to be converted into virtual percepts, as explained earlier in this book. This is done by the perception/response feedback loop, which can be made to support inner speech and facilitate its perception, see Fig. 12.8.

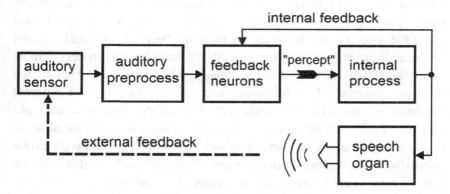

Fig.12.8. The feedback loop model of the perception of the inner speech. Internal feedback allows the virtual hearing of the inner speech produced by the internal process. Otherwise inner speech would have to be spoken aloud in order to become perceived.

Figure 12.8 depicts the feedback loop model of the perception of the inner speech. The internal process produces verbal thoughts that are in the form of neural signals that control the speech organ. A thought that is uttered as audible speech by the speech organ can be heard by the system via the external acoustic feedback loop.

Heard sounds are transformed into a set of neural, auditory feature signals by the auditory pre-process. This process includes the spectrum analysis of the sound. The resulting signals are forwarded to the rest of the system through the feedback neuron group. This neuron group passes normally the auditory feature signals from the auditory pre-process, but it also allows the additional excitation of these signals by the internal feedback signals from the inner neuron groups. The excited signals are auditory feature signals, and consequently lead to the perception of virtual sounds. However, the virtual sounds differ from real sensory sounds in important respects. Firstly, they are not as vivid as real sounds and secondly, their perception is not associated with any activity in the sensory organs and exploratory sensory acts. The turning of the head will not change the apparent direction of virtual sounds. In fact, there is no associated direction for the inner speech.

The feedback loop connections do not have to be designed in. Instead, they may be associatively created by unsupervised learning. This learning could take place in the following way. Let's assume that the inner neuron groups spontaneously generate certain neural signals that cause the speech organ to voice, for instance, "eee". This sound is heard by the auditory sensor, and consequently the auditory pre-process outputs a certain set of auditory feature signals, which are forwarded to the feedback neurons. The feedback neurons receive also the internal feedback signals, which coincide with the auditory feature signals and may now become associated with these. Later on, even with the absence of any external sound, the same feedback signals will associatively evoke internally a similar, but virtual percept of sound "eee". In this way the system may learn to associate a perceived sound with each auditory neural pattern that corresponds to a sound that the sound organ has voiced. Thereafter these imagined sounds may be internally perceived without any overt voicing. The subject will no longer have to talk aloud in order to perceive thoughts in verbal form. (More detailed description in [Haikonen 2007].)

According to the feedback loop model, internal neural firings gain their depiction in terms of qualia when they are fed back to the sensory input neurons, where they evoke the corresponding feature signals. These form the virtual percepts with similar experience that would be evoked by the perception of the real entities.

Sensory perception creates memories; we can remember what we see, hear and experience. In the perception/response feedback loop model, inner speech is virtual perception, and therefore has access to all processes available to sensory perception generally, including the act of memorization. Thus we can remember also our thoughts.

Chapter 13

Examples of Perception/Response
Feedback Loops

13.1. The Auditory Perception/Response Feedback Loop

13.1.1. Basic Requirements

The main function of the auditory perception/response feedback loop is to create a perceived phenomenal sound scene by segregating individual sounds and externalizing their location. Therefore, the design of an auditory perception/response feedback loop should be based on the special nature of sound, the requirements of direct and transparent perception that conserves amodal qualities and the requirement of the externalization of the sound percepts. Auditory preprocessing is needed to satisfy these requirements.

Additional important functions of the auditory perception/response feedback include the prediction of sounds and sound patterns, instant replay memory and the facilitation of the inner speech via auditory introspection. The prediction of sound patterns is related to the recognition of rhythm. The instant replay memory function corresponds to the *echoic memory* of the human brain. The echoic effect extends the apparent presence of transient sounds and in this way helps to focus attention on these even when these are no longer actually present.

13.1.2. Auditory Pre-Processes

The purpose of auditory preprocessing is the production of auditory feature signals that represent the desired information and are suitable for associative processing with distributed signals. Auditory perception is based on the detection of air pressure fluctuations within the audible frequency range, which is usually taken to cover the frequencies from 20 Hz to 20 000 Hz. Sounds with frequencies below 20 Hz are called infrasound, and sounds above 20 000 Hz are called ultrasound. Some animals are able to hear and utilize ultrasound frequencies. Ultrasound frequencies have shorter wavelengths, and allow therefore more accurate sound source direction detection and echolocation. In artificial systems the detectable audio frequency range may be tailored to suit the requirements of each application.

Auditory pre-processing must facilitate *sound segregation*. The ear or a microphone receives the sum of the air pressure variations generated by every sound that is locally present. A microphone transduces this air pressure fluctuation into an analogical temporally varying voltage, see Fig. 13.1.

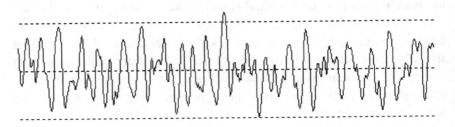

Fig. 13.1. A sound signal from a microphone over a period of time is the cacophonic sum of all incident individual sounds. Individual sounds cannot be segregated without audio spectrum analysis.

Figure 13.1 presents a sound signal from a microphone over a period of time. This signal is the cacophonic sum of all incident sounds, and as such, it does not directly reveal the individual sounds, there is no segregation. Sound segregation, the separation of individual sounds is a necessary condition for content detection and the association of meaning. The sum signal does not tell directly much about the content and there is no way, in which meanings could be associated with the separate sounds of this sum sound. Therefore, a preprocess that would allow the extraction of separate sounds is necessary.

As an auditory pre-process, the inner ear (cochlea) performs a kind of audio spectrum analysis, which allows the grouping of sounds according to their harmonic contents. The use of two ears allow also the detection of the direction of each sound. Artificial sound perception should utilize similar pre-processing methods.

It is known that mathematically every sound can be considered as the sum of sine waves of various frequencies. A periodic sound consists of the sum of sine waves with a fundamental frequency and a number of harmonic frequencies that are multiples of the fundamental frequency. Consequently, a periodic sound can be extracted from a sum of sounds by determining its fundamental frequency and picking up its harmonics. This can be performed, for instance, by the so-called comb filtering. In this way, various sounds can be segregated from each other by the use of frequency analysis.

Spoken language has additional requirements for auditory perception. Heard words must be recognized and learned. This calls for the extraction of audio spectrum, transients and temporal duration. Spoken words are temporally fleeting strings of auditory patterns, called *phonemes* and consequently suitable short-term memories for the capturing and processing of the whole words are required.

The human voice consists of a fundamental frequency and a large number of harmonic frequencies that are multiples of the fundamental frequency. The fundamental voice frequency is also known as the pitch, as it is also the distance between the harmonic frequencies. Pitch determines the overall impression of the voice. Typically, the female pitch is around 200–230 Hz, and the male pitch is around 110–130 Hz. This explains why female voices sound higher.

The intensity of the harmonic frequency components diminishes towards higher frequencies so that the upper frequency limit of the telephone, 3400 Hz, suffices for communication. For a more natural reproduction of speech the upper limit of 5000 Hz would be preferred, and is used in AM radio broadcasts.

We can understand the speech of others regardless of the pitch. In fact, pitch is not very important for the recognition of phonemes, while it may help us to recognize the speaker. Phonemes include vowels (*a, e, i, o, u,* etc.) and consonants (*b, c, d, f, g, h, j, k, l, m, n, p, q, r, s, t, v, w, x, z,* etc.). A spoken vowel is formed, when the spectrum of the harmonic frequencies of the voice is filtered by the mouth and lips. This causes maximums and minimums to the overall spectrum. Mouth cavities are about the same size for every person, and therefore these maximums and minimums for each vowel occur at certain rather fixed frequency bands regardless of the speaker, male or female. The frequency bands for maximums are called formants. Vowels may therefore be recognized by determining the relative energy within each formant frequency band, and these quantities would be suitable auditory features for vowels.

Consonants are characterized by transients, the abrupt changes in sound intensity. The spectrum of a transient sound is rather continuous towards the high end of the frequency range, and does not have a fundamental frequency and its harmonics.

Distributed signal representations require the detection of elementary features. According to the above, useful auditory elementary features may consist of the energies within narrow frequency bands that would cover the whole auditory frequency range. Audio spectrum analysis can be used to resolve these narrow frequency bands and their energies. There are two methods that could be used; these methods are the *Fourier transform* and the use of *filter banks*. The cochlea of the ear is effectively a resonant filter bank.

Sounds and pronounced phonemes may have different durations, therefore the duration of sounds and phonemes must also be determined and represented as a distributed signal array. The detection and representation of sound duration would also allow the representation of rhythm.

13.1.3. A Simple Auditory Perception/Response Feedback Loop

As an example, a simple auditory perception/response feedback loop that includes some of the basic functions is outlined here, see Fig. 13.2.

This auditory perception/response feedback loop utilizes a filter bank for the audio frequency analysis. The filter bank consists of a number of narrow bandwidth filters. The output signals of these filters describe the instantaneous energy of each narrow frequency band, and constitute the auditory feature signals. These signals are forwarded to the feedback neuron group. The output signals from the feedback neuron group constitute the serial stream of auditory percepts that will be broadcast to the other modules of the cognitive system. This serial broadcast preserves amodal rhythm. This property is necessary for any synchronized motor actions including the production of speech.

There are two parallel associative neuron group assemblies within the feedback loop. Both these assemblies learn and reproduce also temporal durations and rhythm. The serial autoassociative memory allows the instant replay of the heard sound as well as the prediction of a sound sequence. The sequence evocation neuron group allows the association of sound sequences with cues and the evocation of these sequences with the same cues. The winner-takes-all (WTA) circuit selects the relevant output.

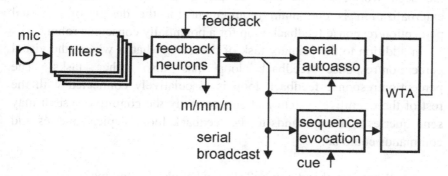

Fig. 13.2. A simple auditory perception/response feedback loop. The sound is analyzed with a filter bank. The intensity values from the filters constitute the auditory feature signals. The serial autoassociative memory allows the instant replay of the heard sound as well as the prediction of a sound sequence. The sequence evocation neuron group allows the association of sound sequences with cues and the evocation of these sequences with the same cues. The winner-takes-all (WTA) circuit selects the relevant output. Serial broadcast preserves amodal rhythm.

During listening to music the feedback tracks and predicts the music in real time. The autoassociative memory predicts the continuation, which will coincide with the actual perceived continuation. It is also possible to focus attention on a specific instrument by the priming effect of the feedback.

13.2. The Visual Perception/Response Feedback Loop

13.2.1. Basic Requirements

The main function of the visual perception/response feedback loop is to tell the cognitive system what is where. In a potentially conscious robot the visual perception/response feedback loop should be able to create the impression of a seen, inspectable external world, and the visual percepts should appear as externalized visual gestalts and qualities.

Visual impressions, including colors, do not have to be similar to the human experience, but they should nevertheless appear as properties of the seen world, for the sake of the grounding of meaning. Thus, externalization is one main requirement for the design of a visual perception/response feedback loop for a potentially conscious robot.

In addition to the primary task of detecting what is where, the visual perception/response feedback loop has also other tasks. The perception/response feedback loop is associatively connected with the rest of the cognitive system and consequently the cognitive system may send queries and commands to the feedback loop. Typical queries and commands could be:

- What is in this direction? (E.g. what was behind me?)
- Where is object x?
- Is this object x?
- Find x!
- Imagine this kind of object!

The visual perception/response feedback loop could broadcast responses like these:

- This is what is here.
- The object x is here.
- The object x has been found.
- This object is or is not the object x.
- The object is moving this way.
- Unexpected action or objects have been detected.

Figure 13.3 illustrates the tasks of the visual perception/response feedback loop.

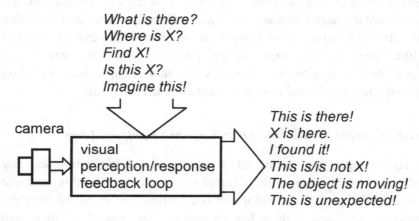

Fig. 13.3. Tasks for the visual perception/response feedback loop.

The successful execution of these tasks calls for the ability to detect objects, the direction where they are and the short-term memorization of these objects and their locations. In addition, the perception/response feedback loop must be able to detect match/mismatch and novelty conditions. The visual perception/response feedback loop is also the enabling machinery for visual imagination and the introspection of the imagined entities.

13.2.2. Visual Pre-Processes

The purpose of visual preprocessing is the production of visual feature signals that represent the desired information and are suitable for associative processing with distributed signals.

Visual perception is based on the detection of the light intensity and wavelength spectrum that are reflected by the observed objects. Advanced visual perception utilizes imaging systems, such as the eye or a digital camera that are able to create optical projections of the environment on a photosensitive surface. However, simpler visual perception systems also exist and are useful.

The visual perception/response feedback loop module shall detect and broadcast visual features and their apparent location. The images from a digital camera consists of pixel maps, which as such do not directly tell much about the seen scene and its objects. Therefore some preprocessing that extracts suitable visual features is necessary. The extracted visual features may include visual change, patterns, colors, the size of the patterns and the motion of the patterns. Optimal visual feature extraction is a challenging task, and cannot be addressed here in detail.

13.2.3. A Simple Visual Perception/Response Feedback Loop

As an example, an outline of a simple visual perception/response feedback loop module with a camera is presented. In this example motion & change, pattern, color and size features are extracted from the camera image. Each of these feature signal groups shall have their own perception/response feedback loops, and thus the complete visual perception/response module consists of the interconnected visual feature perception/response feedback loops, see Fig. 13.4.

In Fig. 13.4 it is assumed that the image sensor of the camera has a small high resolution center area, which determines the focus of visual sensory attention. The direction of the optical axis of the camera is then the gaze direction towards the object. The gaze direction can be changed by turning the camera in pan and tilt directions. A well-defined gaze direction allows good pinpointing of objects and their locations. This is good for the association of meanings and names.

The pan and tilt motors are controlled by the direction perception/response feedback loop, which senses and broadcasts the instantaneous direction of the camera. When the camera pans and scans the visual scene, the instantaneous direction is associated with the instantaneously perceived visual features (connections S2, C2, P2, M2) and in this way short-term memories of the scene are created.

Short-term memory imagery, such as what was previously seen in certain direction, may be evoked by the associative evocation of a direction (connection D1) by an externally applied signal pattern. The evoked direction is returned into a direction broadcast via the feedback loop, and this direction is then broadcast to the other visual feature feedback loops (connections S2, C2, P2, M2). There the direction evokes the visual features that were associated earlier with that direction. The evoked visual features are returned into virtual percepts via the feedback loops. In this way questions like "what was behind me" can be answered.

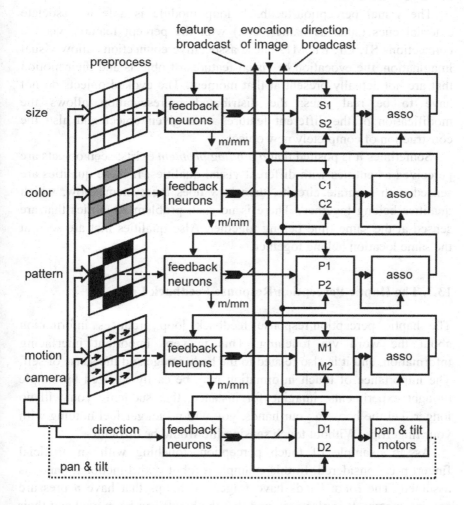

Fig. 13.4. The visual perception/response feedback loop module may consist of a number of sub-modules, such as the pattern, color and motion & change modules. Each of these have their own feedback loops.

The motion & change perception/response feedback loop is also connected to the direction perception/response feedback loop. With the provided visual change information the pan and tilt motors are able to turn the camera towards the visual change. This function allows also the tracking of a moving object.

The visual perception/feedback loop module is able to associate external cues (such as names, etc.) with the percept features via the connections S1, C1, P1, M1. These associative connections allow visual imagination, the evocation of visual features of objects and their motion that are not actually present at that moment. The evoked objects do not have to be real ones; the distributed representation allows the modification of the different features of a given object and also the construction of completely new objects.

Sometimes it is posited that *a binding problem* exists; seen objects are perceived as entities with different visual qualities. If these qualities are sensed with separate circuits, then how does the system know, which qualities belong together? There is no such problem. Qualities that are sensed at the same time belong together. Also qualities that are seen at the same location belong together.

13.3. The Haptic Perception/Response Feedback Loop

The haptic perception/response feedback loop produces information about the word via touching. This is very important interfacing information, and it is also related to the body image and the sense of self. The importance of touch information may be easily realized by simple thought experiments; imagine for instance, that suddenly you will no longer feel the book in your hands, you will no longer feel nothing with your fingertips. Without touch sensing life would be difficult.

As an example of touch perception, touching with an artificial fingertip is considered. In this example a robot with functional hands is assumed. The robot hands have fingers with tips that have a pressure sensing matrix. It is also assumed that the hands can be moved and their position in relation to the body of the robot can be detected by suitable sensors.

When the robot finger is placed on an uneven surface, the pressure sensitive touch sensor matrix produces a coarse "image" of the surface. More information can be acquired by the scanning of the surface by the fingertip. This operation produces two kinds of information. Firstly, it translates the unevenness of the surface into temporal vibrations of the fingertip. Secondly, a series of position-related touch "images" is generated; the touch information comes from the touch sensor matrix, and the position information comes from the hand and finger position sensors. By the combination of these two kinds of information the contours and shapes of larger objects can be felt and their "image" can be constructed with the help of short-term memory.

The touch sensor matrix does not broadcast any location information to the cognitive process. The perceived location of the touch percepts as the point of the fingertip arises from the addition of the position information of the hand and fingers. This information can originate from the joint position sensors and can also be achieved visually, if the robot is sighted. In this way the touch percepts are externalized to the tips of fingers and a contribution to the body image is made.

According to the aforesaid, touch perception involves the cooperation of at least two sensory modalities, namely the touch modality and the body position (proprioception) modality. This can be achieved by the associative interconnection of the related perception/response feedback loops, see Fig. 13.5.

In Fig. 13.5 the touch sensor matrix delivers pressure-related signals from each individual pressure sensor to the feedback neurons. The pattern of these signals is a kind of a coarse image that represents the structure of the sensed surface.

The position perception/response feedback loop delivers the corresponding location of the fingertip. The position of the fingertip is sensed by a variety of proprioceptive sensors that detect the angles of each joint in the hand-finger mechanical system. The resulting percepts define the position of the fingertip in three dimensions.

The sensed touch pattern and its location are associated with each other by the short-term memories STM1 and STM2 that keep track of what is currently where.

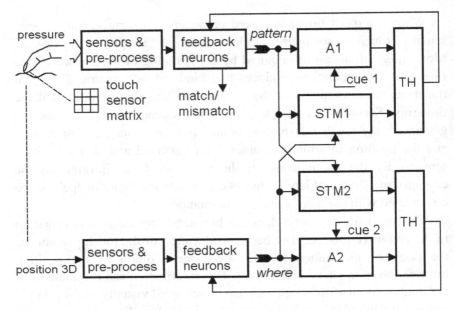

Fig. 13.5. The perception/response feedback loops for touch perception.

The associative neuron groups A1 and A2 are non-volatile associative memories. The neuron group A1 associates names or some other cues with learned touch patterns, and the neuron group A2 associates similarly cues with positions.

Thus, a position cue, "imagined position", at the neuron group A2 will evoke the corresponding position signals in the position perception/response feedback loop, and may lead to the positioning of the hand and finger to that position. This position information, "where", is also forwarded to the neuron group STM1, where it will evoke the latest touch percept at that position. This is forwarded to the feedback neurons via the feedback loop, and will now become an expectation for the touch percept. Match and mismatch will be detected.

When an object is felt by scanning its contours by fingers, more or less complete information about the shape of the object is accumulated in the short-term memories. This information will allow, for instance, the manipulation of objects in darkness.

Chapter 14

Symbols in Perception/Response Feedback Loops

14.1. From Sub-Symbols to Symbols

A perception/response feedback loop operates inherently with sub-symbolic representations, patterns of sensory signals. These signals indicate directly presence of the corresponding sensorily perceived or introspected features and their combinations, but apart from that they do not have other inherent meanings. These kinds of direct sub-symbolic signal patterns are self-explanatory and a necessary prerequisite for the grounding of meaning, and must therefore be utilized in cognitive machines that understand. However, higher cognition calls also for symbolic presentations, as language and abstract thinking generally use symbols. Symbols are necessary, but what would be their origin and how could they be inserted into a sub-symbolic neural system?

All signal patterns that circulate in perception/response feedback loops are fundamentally sub-symbolic sensory signal patterns. Symbols in perception/response feedback loops would also have to have the same form, as nothing else is available or can be inserted. But sub-symbolic signal patterns are not innate symbols; true symbols depict entities that are not inherently related to them or present in their appearance.

However, here is the trick. The meanings of symbols are not inherent, they must be learned. Anything can be used as a symbol, if a meaning can be associated with it. Therefore sub-symbolic sensory signal patterns in perception/response feedback loops can be used as symbols by associating a meaning with them.

14.2. The Association of Meaning

Percepts may have associated meanings. A tool is not only a certain perceived object. In addition to its appearance, its percept evokes also things that are related to it and associated with it; its purpose and possibilities for its use.

However, a percept may also be made to be associated with something that is not related to the percept naturally. This allows the use of a percept as a symbol, by associating a meaning with it. In this way it can be used to stand for matters that are not necessarily a property or appearance of the actual sensory stimulus. These kinds of associated meanings require neural connection mechanisms that allow the association and evocation of percepts from various modules with each other. The previously described perception/response feedback loops operate associatively, and can be cross-connected to provide this function.

As an example, the principle of the association of a spoken word with a visual pattern is presented in Fig. 14.1. This operation is done via associative cross-connections between perception/response feedback loops of the visual and auditory modules.

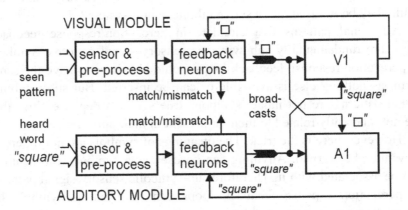

Fig. 14.1. Associative cross-coupling of modules allows the association of meaning. Here two modules, visual and auditory, are associatively cross-connected. In this example the word "square" is associated with the seen square pattern. When the visual module detects a square pattern, the name "square" is evoked in the auditory module and vice versa.

In Fig. 14.1 the visual module has an image sensor that is able to produce images of external objects. The image information is preprocessed, and a number of visual feature signals are forwarded to the feedback neuron groups. The auditory module is able to detect spoken words, which are processed into a number of feature signals. Without any associative cross-connections visual percepts would remain as visual experiences without any further meanings and heard words would remain as meaningless sound patterns.

The cross-connection lines between the two modules enable the formation of associative cross-connections. These connections are formed by supervised or unsupervised learning, and will be activated by the presence of the auditory or visual percept that is associated with a percept at the other modality.

During associative learning, the percept from the visual module is broadcast to the associative input of the auditory module neuron groups, and vice versa.

For example, let us assume that the visual module is perceiving a square, and at the same time the auditory module is perceiving the word "square". At this moment, if learning is enabled, the visual neuron groups V1 will associate the visual percept of square with the auditory percept of "square", and the auditory neuron groups A1 will associate the auditory percept of "square" with the visual percept of square. As the result, simple two-way labeling has taken place.

After learning, the word "square" will act as a symbol for the visual percept of square, and it will evoke the percept of square at the visual module also in the absence of any visually perceived external square. On the other hand, if a square is shown to the system, the visual percept of square will evoke the word percept "square". Now both the visual square and the word "square" have an associated meaning. This simple two-way action is the fundamental enabling factor for symbolic processing in the associative neural network by allowing the generation of grounded symbols in the first place. During this action, the two modules are focusing on the same entity, albeit in their own terms, and in this way they have formed a coalition.

This kind of coalition of two modules allows also other meaning-related functions. For instance, if in the previous example the word "square" and the actual visual square were presented at the same time, match-conditions would occur at the feedback neurons. Thus, with the association of the words "yes" and "no" with the match and mismatch conditions, the system would be able to answer "yes" to the question: "Is this a square?". If the presented word and the visual object do not match, then mismatch conditions occur, and the system will answer "no" [Haikonen 1999].

Simple two-way labeling of percepts does not yet provide much in the way of cognition, but the situation changes, when the basic associative mechanism is applied to a system of many sensory and motor modules with memory, using also the association of temporal sequences and memories. In this way the meaning of percepts will not be limited to mere labels. Instead, the associated meanings may be many, including sensed and executable motor actions. The activation and selection of the associated meanings would depend also on context and the instantaneous emotional situation. Also, strings of evoked virtual percepts might follow, as each evoked virtual percept would evoke its own associated meanings. These learned meanings would constitute an associative network of experience that the system would be able to utilize for its benefit in each situation. This is information integration over multiple modules, and it is the key to higher cognition. Examples of information integration with multiple modules are given in the next chapter.

Chapter 15

Information Integration with Multiple Modules

15.1. Cooperation and Interaction of Multiple Modules

Higher cognition goes beyond simple two-way labeling and naming, and requires more advanced information integration and interaction between several modules. This can be achieved by additional associative cross-connections between the modules.

As an example of the associative cross-connection and interaction between multiple modules an extension of the previous principles is presented. In this example the cooperation and associative interaction of four modules during an object searching task is described. The modules are the linguistic auditory module, the visual pattern module, the visual color module and the motor module. The relevant associative interconnections of these modules are given in Fig. 15.1.

The execution of the given task begins, when the auditory module receives the verbal request *"find cherry"*. These words are broadcast to the visual and motor modules. The word *"find"* is associated with searching at the motor module, and accordingly it will trigger a search routine, such as the turning of the head. The motor module does not have to know what is being searched for, and consequently the word *"cherry"* is not associated with any motor action and is ignored.

The visual modules ignore the word *"find"*, because it is not associated with any visual object. The word *"cherry"* is associated with a pattern and a color and will evoke the pattern feature *<round>* at the visual pattern module and the color feature *<red>* at the visual color module. These are fed back to the corresponding feedback neuron groups P and C.

When a cherry is found and seen, the sensory features of <*round*> and <*red*> match the feedback features, and match signals are generated. A *match*-signal is also broadcast to the motor modality, which now terminates the search operation leaving the perceived object in the focus of visual attention.

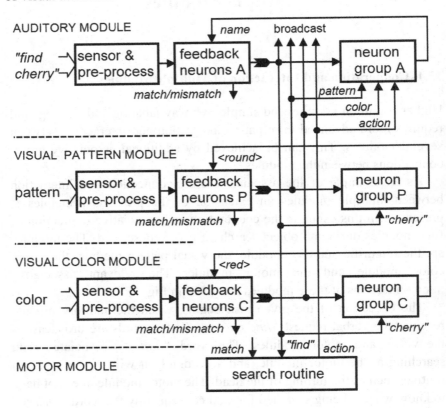

Fig. 15.1. The associative cross-connection of four modules. When the auditory module receives the request *"find cherry"*, the word "cherry" evokes the virtual features <*round*> and <*red*>, which are compared with the pattern and color percepts. When the evoked features match the perceived features, the object has been found, and the match-signal terminates the search action.

The pattern and color percepts are also broadcast to the auditory module, where the names of the pattern and color may be evoked, but also their combinations may evoke the name of the corresponding object, if it has been previously named. In this case the word "cherry" might be evoked. The *action* signal pattern from the motor module is broadcast to the auditory module and this interconnection allows the verbal naming and reporting of the on-going motor activity.

The network of associative cross-connections in this example can be easily extended to systems with a larger number of modules based on the perception/response feedback loop principle.

15.2. Sensorimotor Integration

It was presented earlier that conscious perception of the environment involves the externalization of the internal phenomenal appearance of the related neural activity. Due to this externalization the world appears to be out there, outside our brain and body. This outside-the-body externalization effect is present with the non-contact senses of vision and hearing. Externalization applies also to body sensations making them to appear outside the brain.

The externalization of the appearance of sensory information allows seamless sensorimotor integration, the direct physical interaction with the environment, because the externalization adds the sense of direction and distance to the perceived object. For example, when we see a book on a table, we can reach out and pick it up without any conscious symbolic computations for the necessary hand movements. The perception of the external position of objects within our reach is seamlessly coupled to the neural circuits that control the motion of our hands and body, allowing thus the possibility for direct motor actions. The readiness for these actions arise automatically as soon as objects are perceived, without any special preparations; the command to execute the act would arise from the cognitive context.

Traditional robots do not have this kind of seamless sensorimotor integration. There is no subjective experience with an internal appearance at all, and consequently no externalization effects can take place; there is nothing to be externalized. The digitalization and symbolic processing of sensory information effectively voids this possibility and the robot is reduced to a blind mechanism. Every motion that a traditional robot makes, must be computed by preprogrammed rules. If a robot were to pick up a book, it would first have to determine the exact position of the book, and next it would have to compute a suitable trajectory for its hand. During the execution of this task the robot would have to determine the exact position of its hand, and compare this with the computed trajectory. Any error in the instantaneous hand position would have to be corrected. True enough, this can be easily executed by digital signal processing and well-known feedback control loops, but nevertheless, preprogrammed algorithms for each executable motor routine, action and action sequence would be necessary. Computers do not have any "ex tempore" action modes and cannot improvise any new action routines on the fly.

Associative information processing offers another, direct way for the realization of sensory percept externalization and sensorimotor integration, without any computations or preprogrammed motor routines. However, associative robots would still use conventional electric motors, which do not have associative interfaces. Therefore, the practical design of associative sensorimotor integration calls for the design of interfaces that allow the control of motors by associative distributed signal representations. In the following, the traditional motor feedback control loop is explained first, and then its connection to an associative perception/response feedback loop is described. Next, it is explained how this arrangement enables seamless sensorimotor integration.

15.3. Feedback Control Loops

A typical effector system consists of a motor that moves some moving part, for instance, a robot hand. The instantaneous position of this moving part is detected by a position sensor that outputs a signal that is proportional to the sensed position. The system is controlled by an input signal that indicates the desired position for the moving part. The difference between the desired position (set-value) and the sensed position (is-value) is computed (in analog systems no numeric computations take place, the difference can be determined by simple resistor circuits). If the difference between the sensed position and the desired position is zero, then the moving part is at the desired position. If the difference is not zero, then the polarity of the difference signal indicates the direction in which the motor must run, so that the moving part can achieve the desired position. This system arrangement is called a feedback control loop, see Fig. 15.2.

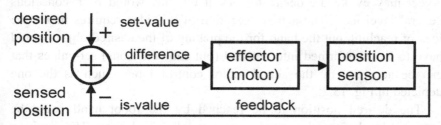

Fig. 15.2. The principle of the feedback control loop for a motor (effector control loop). The effector moves a moving part whose exact position is measured continuously by the position sensor. The difference of the desired position (set-value) and the measured position (is-value) is continuously computed, and this difference drives the effector. When the difference is zero, the desired position has been reached and the effector stops.

Figure 15.2 depicts a simple proportional feedback controller. This arrangement works, but is not very accurate. Obviously, when the desired position and the sensed position are very close, the difference signal will become very small and the motor may no longer have a sufficient drive signal. This leads to a small steady-state position error for the moving part. In actual industrial applications proportional-integral-derivative controllers (PID controller) are often used, because their steady-state error is zero. For the purposes of this discussion, the proportional feedback controller model is sufficient.

In order to make a feedback controlled effector to execute a desired motion, two signals must be provided, namely the desired position signal (set value) and the signal that indicates the actual sensed position (is-value). In analog systems continuous signals (voltages) are usually used; the amplitude of these signals is proportional to the position values.

15.4. Hierarchical Control Loops

The cognitive control of motor actions is hierarchical. The actual desire to execute a motor action, such as the grabbing of an object arises at circuits that are not directly related to the actual motor circuitry. A seen object may evoke the desire to pick it up; this would be a conscious general level idea. The actual execution of this idea requires more; the idea of reaching out the hand for the picking of the desired object would have to be transformed into detailed (but sub-conscious) set-values that can be accepted by the actual motor control loops, such as the one depicted in Fig. 15.2.

The desired position to be reached by the robot hand could be provided by the visual perception/response feedback loop. This position might be the position of a seen real object or it might be an imagined position. The actual hand position information would be provided by the perception/response feedback loop that senses the relative positions of the moving body parts (proprioception). However, the position representations of these two perception/response feedback loops would not be compatible with each other. The visual position percepts and the proprioception percepts would be represented in terms of their corresponding sensors, and therefore the resulting signal patterns would not have one-to-one signal-wise correspondence. Therefore the difference between the sensed position and the desired position cannot be determined directly from these signal patterns. This compatibility problem can be overcome via hierarchical associative connections.

Figure 15.3 depicts the hierarchical control mode of action that is effected by the cross-connection of the visual perception/response feedback loop and the proprioception perception/response feedback loop. The desired visual position is broadcast to the associative neuron group at the proprioception loop, where it evokes the corresponding position representation in proprioception terms. The actual motor action is executed by the effector loop, which is similar to that of Fig. 15.2. The effector loop will run the motors until the proprioceptive is-value matches the set-value.

In Fig. 15.3 proprioceptive sensors detect the current position of the moving part and forward this in information in the form of distributed signal representation to the feedback neurons. Normally, this representation would also be the percept representation, which would be broadcast and also forwarded to the inner associative neuron group. The output from the proprioception loop's associative neuron group would be the set-value, and the output from the proprioceptive sensor would be the is-value for the effector loop.

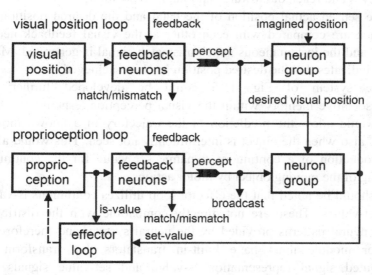

Fig. 15.3.The associative motor control system based on a perception/response feedback loop. Proprioception gives the current position of each body part (is-value). The desired visual position evokes the corresponding set-value for the effector. The effector loop has built-in translators that transform the distributed signal representation is-value and set-value signals into continuous signals so that their difference can be determined. The proprioception feedback loop allows also the virtual execution of motor actions.

In Fig. 15.3 both the desired position and the actual position information are in the form of distributed signals. The proprioception and visual perception systems are separate modalities, and therefore their distributed signal representations depict dissimilar quantities; there is no common code. Therefore the desired position distributed signal representation must be translated into the quantities that are used by the proprioception loop. This is done by neuron groups via associative learning. During the initial learning phase the system of Fig. 15.3 would visually perceive various positions of the moving part, and these position representations would be then associated with the corresponding proprioceptive representations at the proprioception loop. Thereafter a visual position percept would be able to evoke the corresponding proprioceptive representation. It should be noted that the desired position does not have to be necessarily seen, it may also be imagined (one may close eyes and reach out towards an imagined position).

The sensed actual position of the hand and the desired position via feedback are compared with each other at the visual feedback neuron group, and the instantaneous match/mismatch signal is generated. Match signal indicates that the desired position has been achieved.

The system of Fig. 15.3 could be developed further. An autoassociative memory within the visual perception/response feedback loop would allow the prediction of the trajectory of a known moving object, also when the object is intermittently not seen. This would allow the production of a continuous dynamic set value for the continuous tracking of the moving object by motor actions.

It should be noted that the effector loop utilizes continuous is-values and set-values. These are not directly compatible with the distributed signal neural patterns provided by the neural system, and therefore the effector module must have built-in translators that transform the distributed signal representation is-value and set-value signals into continuous signals. (See [Haikonen 2007]).

Practical effector systems, such as robot hands, would have several joints and motors. Every joint would have position sensors and the motors would have an associative control loop.

Associative seamless sensorimotor integration of the visual modality and the motor modality allows easy physical interaction between a robot and the environment without explicit computations. Sensorimotor integration is also one contributing factor to the externalization of non-contact percepts; reaching out for an object means that the object is out there.

Chapter 16

Emotional Significance in Associative Processing

16.1. The Significance of Percepts

In principle, an associative information processing system is able to accumulate a large number of possible associative connections between entities. Any stimulus would activate many associative links, and as a result, many signal patterns would be evoked. This leads to the problem of choice, because the evocation of all possible associations would overflow the system. This is a serious issue that is known also as the *combinatorial explosion* problem. It is also related to the so-called *frame problem* [Dennett 1987], the question of which associated meanings would be the most relevant in the framework of each situation. Therefore the scope of choice must be limited in one way or another. In associative information processing, the associative evocation process must be guided in such a way that at each moment only the most relevant associative connections would be activated. This calls for mechanisms that are able to guide attention and pinpoint the most important associations. Context is one such mechanism that is easily realizable in associative systems, but it is not sufficient in every situation.

An example provided by Dennett [1987] illustrates the problem of choice: A robot enters a room, in order to find and retrieve a certain object from a large selection of various objects. Among these objects is a bomb, and its fuse is burning. What should the robot do? Should the robot go there and start looking for the object to be retrieved? This might take a while, and in the mean time the bomb would explode. Obviously the robot would have to focus its attention on the bomb immediately. However, how could this happen, if the robot were guided by context only, which in this case would be related only to the object to be

149

retrieved? Generally, context would frame and limit the scope of choice and should be used as an attention guiding factor, but it works against reality in situations like this. A context-guided robot would be looking for a certain object and while doing so, it would ignore all the other items that do not fit the context and do not resemble the searched object. In doing so, the robot would also ignore the bomb, because in the robot's mind the bomb would have the same irrelevance as all the other unrelated items. Ignoring the bomb would be a stupid thing to do, but there is nothing in the context that would alert the robot about the immediate danger.

Obviously, in addition to the context, there is a need for a mechanism that would be able to evaluate instantly the significance of each perceived object and situation. If necessary, this evaluation should be able to override the context-related attention and focus attention on the more significant percepts. Emotional significance would do this.

16.2. Emotional Significance of Percepts

Emotional value systems are mechanisms for the association of importance and significance with sensory percepts and percepts of mental content. These systems learn and associate emotional significance values with percepts, and use these values to focus attention on the most important ones. These systems can also work as motivational drivers; good is to be pursued and bad is to be avoided.

Emotional value systems may use the experiences of pain and pleasure for the grounding of good/bad emotional values. A simple emotional value system may consist of pleasure and pain sensors and additional associative neuron groups. These neuron groups would be able to associate pain and pleasure signals with various percepts from other sensory modalities. This arrangement would use pleasure and pain signals for the evocation of various system reactions and the amplification of broadcast percepts with strong emotional significance. A simple emotional value system is depicted in Fig. 16.1.

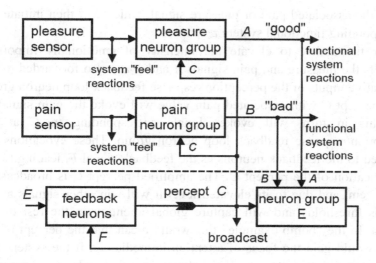

Fig. 16.1. A simple emotional value system and its connection to a perception/response feedback loop. Percepts are associated with simultaneously occurring pain or pleasure, and later on will evoke the associated pain or pleasure signal. This will focus attention on the percept. System "feel" reactions convey the feel and may include withdrawal etc.

The simple emotional value system of Fig. 16.1 consists of pleasure and pain sensors and pleasure and pain neuron groups. Pleasure and pain sensors generate neural signals that evoke directly their typical system "feel" reactions, such as the effects on attention via threshold modulation. These system reactions are not symbolic, and they would correspond to the feel of the perceived pleasure and pain. Pleasure and pain signals evoke also functional system reactions and responses, which may be learned.

The operation of the simple emotional value system of Fig. 16.1 is twofold. First, it associates pleasure and pain with percepts giving them good/bad emotional values. Secondly, it elevates the intensity of emotionally important percepts.

In Fig. 16.1 the pleasure neuron group associates a sensory percept *C* with pleasure *A,* if pleasure occurs simultaneously with the percept *C* of an event *E*. Likewise, the pain neuron group associates pain with a percept *C*. As soon as these associations are learned, the percept *C* will evoke the associated pain or pleasure signal, which will then initiate the corresponding functional system responses.

For the ability to elevate the intensity of emotionally important percepts the pleasure and pain signals A and B are also forwarded to the associative inputs of the perception/response feedback loop neuron group E. A percept C with associated pain value will evoke the pain signal B, and this in turn, will evoke signals corresponding to C at the perception/response feedback loop neuron group. These evocations are returned to the feedback neurons as the feedback signal F, leading to the amplification of the percept C. The amplified percept C is broadcast to the system, and due to its elevated level it will win other signals at the various thresholds and will capture global attention in the rest of the system. In the "bomb example" this would mean that the percept of the bomb would gain the focus of attention immediately, if the system had earlier learned the emotional significance of a bomb. As a parallel process this evaluation is immediate.

The system principle of Fig. 16.1 allows also the emotional good/bad evaluation of internally evoked virtual percepts, such as the products of imagination. This function is useful in the good/bad evaluation of the outcomes of planned and imagined actions.

Chapter 17

The Haikonen Cognitive Architecture (HCA)

17.1. Cognitive Architectures

Cognitive architectures are system arrangements for the production of human-like cognition. Cognitive architectures try to combine and implement a large number of cognitive functions within a single system, and may be considered as functional models for the biological brain or as blueprints for robot brains. Typical cognitive functions that may be included are:

- Perception
- Attention
- Prediction
- Learning
- Memory
- Imagination
- Planning
- Judgment
- Reasoning
- General intelligence
- Emotions
- Natural language
- Motor action control

Early Artificial Intelligence approaches implemented only one or few aspects of cognition and consequently were quite fragile; the approach failed outside the narrow scope of its competence. The targeted general intelligence was not, and has still not been achieved. Cognitive approaches try to remedy this by integrating many cognitive functions in a system, and in this way try to achieve a more universal cognitive competence, which would be comparable to that of humans.

An advanced cognitive architecture should also address the issues of consciousness, self-consciousness, subjective experience and qualia. A robot with an implemented cognitive architecture should produce human-like responses, action and behavior. This robot should know and understand what it is doing, and it should be able to act autonomously in meaningful ways in new situations. It should also be able to learn and accumulate experience, and in this way become more proficient and skillful during the course of time. Cognitive architectures may be based on different approaches.

The high-level symbolic approach defines the action of the implemented cognitive functions directly, and may assign a dedicated system block for each cognitive function. These function blocks are then logically connected to each other. This approach is usually implemented as a computer program, and represents typical symbolic AI tradition.

The low-level sub-symbolic approach begins with the definition of the function of fundamental components such as the neuron and synapse and their computational algorithms. Next, a multilayer network of interconnected neurons is devised. This network would then compute the desired cognitive function. This approach is not well suited for symbolic processing, and therefore hybrid systems that combine symbolic digital computing and neural networks have been developed. The low-level sub-symbolic approach is the traditional artificial neural network approach.

The associative sub-symbolic/symbolic neural networks approach begins also with the definition of the function of the fundamental components such as the neuron and synapse and their artificial implementation. In this case the neurons are different from those of the traditional artificial neurons. The neurons and neuron groups used in this approach are associative, and they allow the association of signals and signal patterns with each other in a direct way. This approach allows symbol grounding via self-explanatory information, the association of meaning and the seamless sub-symbolic/symbolic operation.

A cognitive architecture may include a master executive and/or a master attention control. In other cognitive architectures these operations may be distributed along the system.

Cognitive architectures are often described with the aid of block diagrams. In engineering, block diagrams are very useful, when a general view of a complicated system must be presented. The blocks of a block diagram represent functional modules, and the connecting lines between the blocks represent the information flow and functional connections between the modules. A block diagram is a valid presentation only if the inputs, outputs and the executed functions inside the blocks are defined. In software systems a block represents a program routine, and in electronic hardware realizations a block represents a circuit diagram. A circuit diagram (schematic diagram) depicts the actual connection of electronic components such as transistors, resistors and capacitors for the execution of the desired function. A block may have several alternative ways of implementation. A block diagram without well-defined blocks and interconnections has only entertainment value.

A large number of cognitive architectures have been developed and proposed during the last decades. For instance, Samsonovich [2010] lists up to 26 architectures in a catalog of implemented cognitive architectures, and still this list is not complete. Duch *et al.* [2008] give a compact review of the different approaches of cognitive architectures.

The author has designed a cognitive architecture that combines the principles of neural associative information processing described in the previous chapters. This architecture is known as the Haikonen Cognitive Architecture (HCA) [Haikonen 2003, 2007, 2012] .

17.2. General Overview of HCA

The Haikonen Cognitive Architecture (HCA) combines the afore presented principles and circuit solutions into a dynamic system for the production of possibly conscious human-like cognition with meaning including the flow of inner imagery and natural language inner speech.

As previously noted, processing with meaning requires the solving of the symbol grounding problem. This can be achieved by using self-explanatory information, which has the form of qualia. The presence of qualia is the main hallmark of phenomenal consciousness, and successful implementation of qualia leads to conscious machines. The HCA is designed around a sub-symbolic perceptual system, which allows the acquisition of sensory information in self-explanatory forms for the purposes of symbol grounding. The perception processes of the HCA also allow percept externalization with the help of seamless multisensory and sensorimotor information integration. The HCA uses associative information processing with distributed signal representations, and utilizes emotional significance evaluation and match/mismatch/novelty detection as attention and relevance guiding factors. Information processing is executed by associative neural networks that allow the transition from sub-symbolic to symbolic processing.

Figure 17.1 depicts the generalized units and the information flow paths between them in the HCA.

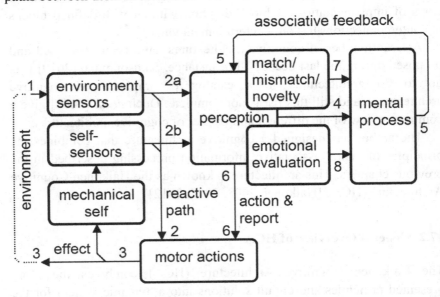

Fig. 17.1. The general principle of the information flow in the Haikonen Cognitive Architecture (HCA). The operation is based on perception process, which allows also the introspection of the results of mental processes as virtual percepts via the associative feedback. Note; speech is also a motor act.

In Fig. 17.1 the environment and the mechanical self (the body and its functions) are sensed by a number of sensors. These sensors output their signals to the perception process and also directly to the motor action unit. The perception process is coupled to emotional evaluation and match/mismatch/novelty detection. They all output their signals and effects to the mental process. The output of the mental process is fed back to the perception process, so that the requirements of priming, prediction and introspection are fulfilled.

The general configuration of Fig. 17.1 allows the following modes of operation:

1. *Reflex reaction.* Stimulus → reflex (1 → 2 → 3). Here the stimuli 2a, or 2b evoke directly motor responses. This corresponds to the stimulus-response reflex, for instance, the withdrawal of the hand from a flame. This mode of operation does not produce conscious reports of what is happening; reports may be generated afterwards, when the perception process and global attention becomes involved.

2. *Sub-conscious routines.* Stimulus → percept → action (1 → 2 → 6 → 3). This corresponds, for instance, to the automatic manipulation of a small object just because it is within reach, without any real need or global attention focus. This action can be reported if global attention is focused on it.

3. *Deliberated actions.* Stimulus → evaluated percept → associative deliberation → evaluated introspective percept → action (1 → 2 → 4,7,8 → 5 → 6 → 3). This signal path corresponds to a deliberated loop; the motor response arises as the result of perception, evaluation and planning. This action can be reported.

4. *Imagination.* Evaluated introspective percept → associative deliberation → evaluated introspective percept → associative deliberation → ... (4,7,8 → 5 → 4,7,8 → 5 →...). This loop would correspond to imagination and free-running thoughts, where the products of the mental process are perceived and emotionally evaluated, as if they were sensory percepts. These

virtual percepts may then be used as the cues for the subsequent thoughts. The results of this process can be reported.

These basic modes of operation of the HCA satisfy the requirements of cognition on a general level. In the following it is described, how these functions are realized with an architecture consisting of a number of cross-connected perception/response feedback loop modules.

The operation of the HCA can be simulated by computer programs, but in this case the sub-symbolic function is lost, the perception process will not be direct and no symbol grounding, qualia or perceptual externalization effects will be present.

17.3. The Block Diagram of HCA

The complete HCA consists of a number of associatively cross-connected modules that utilize the perception/response feedback loop principle. A simplified block diagram depicting the organization of the complete HCA is given in Fig. 17.2.

The block diagram of Fig. 17.2 depicts the HCA with seven modules. The number of modules is not fixed to the seven shown here, depending on the application more or fewer modules may be included. The HCA block diagram not a computer program flowchart, instead it depicts a dynamic integrated system with parallel and temporal processes and reactions.

Sensory information is preprocessed into feature signals. Each feature signal has its own perception/response feedback loop. Thus, each module consists of a large number of parallel single signal feedback loops. Each percept consists of an pattern of feature signals. Apart from sensors and effectors, the blocks depict associative neuron groups.

In Fig. 17.2 the module 0 consists of internal need detectors such as energy levels etc. (In humans: hunger, thirst, tiredness, etc.). This module generates basic motives for action. The module 1 contains pain and pleasure sensors, and acts as emotional evaluator and motivator.

The modules 2, 3 and 4 are sensory modules for touch, vision and sound perception.

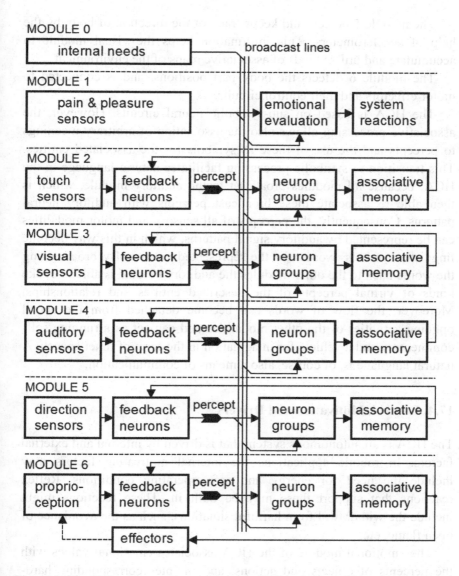

Fig. 17.2. The Haikonen cognitive architecture (HCA) consists of perception/response feedback loop modules that broadcast their percepts to each other. Each module consists of a large number of parallel single signal feedback loops. Each percept consists of a pattern of feature signals. Interconnecting lines depict paths for signal patterns. Apart from sensors and effectors the blocks depict associative neuron groups. Threshold control interconnections are not indicated here.

The module 5 detects and keeps track of the direction of body by the help of accelerometers. This information helps the visual module to accumulate and utilize kinds of associative maps of the environment.

The module 6 detects the body part positions, and is connected to motor effectors and their control circuits.

The HCA is based on sub-symbolic neural circuits. However, the associative processing allows also the association of arbitrary meanings to percepts and the use of these as symbols for the associated entities. This transition to symbolic processing facilitates natural language in the HCA, because all modules broadcast to the auditory module, which is then able to associate these broadcast percepts with auditory neural patterns. Consequently, the percepts of all sensory and motor modalities can be represented by auditory signal patterns, which in this way become linguistic symbols, words, for the represented entities. By broadcasting the words back to the other modules the auditory module is able to evoke kinds of virtual percepts of the described entities and relationships. Moreover, the flow of words can become detached from the actual ongoing activities of the other modalities and can act as a running self-commentary and a vehicle for imagination in the form of inner speech. A natural language is, of course, also a means of communication.

17.4. Control, Motivation and Drivers

The HCA is an autonomous system that is driven by internal and external factors. In robotic applications the internal motivating factors may include the need for energy and safe operating conditions. Reflex reactions that support these may be built in. These reactions might include the withdrawal from harmful situations, such as the avoidance of open flame, etc.

The emotional module of the HCA associates emotional values with the percepts of objects and actions, and initiates corresponding hard-wired system reactions. The hard-wired system reactions of pain and pleasure are important motivators. Actions associated with pain would be avoided and actions associated with pleasure would be sought for.

External association of pleasure and pain as reward and punishment can be used for motivation.

External factors, opportunities, provided by the environment may also motivate and trigger action. An observed opportunity to execute an action would initiate imagined execution of this action. This imagined action would be transformed into real action, if no counteracting factors were present. These counteracting factors would include the outcome of emotional evaluation of the imagined action and any external situation-related factors.

External opportunities as action triggers could also be allowed to lead to the playing and toying with objects. This would facilitate the unsupervised learning of the ways of the external world.

17.5. Information Integration, Coalitions and Consciousness

Multisensory and sensorimotor information integration is necessary for cognition. Percepts from different sensory modalities must amount to a coherent view of the world, and this view must allow correct motor actions. The percepts must be able to evoke meanings and affordances, and their emotional significance must be evaluated.

Sensorimotor integration is also required for the externalization of percepts, which would allow the creation of the impression of the observable world out there and also the acquisition of the body image. All this calls for seamless cooperation and coalition of the modules and associative memories. This kind of coalition results in the activation of large number of associative paths. This in turn facilitates the forming of associative memories and the possibility to report experienced episodes later on.

Reportability results from information integration, and is a hallmark of consciousness. Therefore information integration and the coalitions of modules are considered by some researchers as necessary prerequisites for consciousness. According to this view consciousness arises when all modules coalesce and focus on the same entity; one unity, one consciousness [Tononi *et al.* 1998, Shanahan 2010].

The HCA fulfills the requirements of seamless multisensory and sensorimotor information integration and percept externalization by the associative interconnections between the modules (broadcast). However, the HCA allows also partial and concurrent coalitions of modules.

At first sight it might appear that concurrent coalitions are not needed, and are actually harmful, as they might destroy the global unified focus of consciousness. This is not so, and can be seen, for instance, from the act of reading: You are now reading this page, this line and right now this word. Now you are reading the words that follow. Your eyes scan the text line by line, word by word and every now and then your hand turns the page. It is your brain that controls these operations, yet you are not really aware of these, and why should you. After all, these operations do not contribute anything to the content of the text that you are reading. Nevertheless, the brain must provide resources also for these operations; sensory resources to see the text and motor resources to control the eyes and your hand when you turn the page. When these actions are considered in terms of module-based cognitive architectures, it will be noted that all these actions require coalitions of several modules. In addition, these coalitions take place concurrently with those coalitions that allow the understanding the text. Thus the whole operation of book reading recruit concurrent coalitions of modules. In a similar way, most tasks that we do, call for concurrent coalitions of modules, and the activity of many of these coalitions remain sub-conscious. It should also be understood that sensory modalities, like vision, may consist of a number of sub-modules, which may individually take part in different coalitions.

As a distributed architecture with extensive cross-connections, HCA allows easily the formation of ad hoc coalitions that may be needed in each situation. Only the action of those coalitions that are able to produce report and memory percepts will become consciously perceived. This fulfills the reportability prerequisite for consciousness. The first and mandatory prerequisite for consciousness, the inner phenomenal appearance of information, is addressed by the direct and transparent perception process and the introspection via feedback loops as discussed before.

Chapter 18

Mind Reading with HCA

18.1. Mind Reading Possible?

Would it be possible to develop technologies that would allow the reading and viewing other people's thoughts and inner imagery and perhaps recording these as a video? This kind of technology could have important applications in the diagnosis and care of brain injuries and especially in the care of *locked-in syndrome* patients. A locked-in syndrome patient is awake and aware, but is not able to move or communicate verbally due to the paralysis of most of the voluntary muscles. In addition to patient care, mental content imaging technology would also open a new window for brain research.

We experience our mental content in the form of qualia. It was argued earlier that qualia are not directly accessible to other people via any brain imaging technologies, as these can only detect and measure physical processes that are related to the brain activity, and produce *symbolic representations* of these. Qualia are direct ways of information representation, while symbolic representations are just descriptions. Brain imaging methods do not detect qualia or mental content, instead they detect physical processes that may or may not be related to conscious mental content. Thus, the direct detection and recording of mental content would seem to be impossible.

However, it was also argued before that some of the brain activity constitutes the conscious mental content and the qualia are the way, in which the conscious brain activity appears to the subject. This, if true, would allow an indirect method for the imaging of mental content, because technologies for the detection and measurement of brain activity exist. Vice versa, the successful imaging of mental content would prove the equivalence of the physical brain processes and the mental content (and would also justify the main tenets of this book). This hypothesis has the following consequence: The data, the symbolic representations of the physical processes produced by various brain imaging technologies, would be related to the mental content, and this content could be determined from this data. The crucial question of technological mind reading is: What kind of method could extract mental content from the data produced by the various brain imaging technologies? Surprisingly, the principle of this kind of a method is a simple one.

18.2. The Principle of the Imaging of Inner Imagery

According to the perception/response feedback loop hypothesis, visual perception and internal imagination produce neural activities that occupy the same neural area. In the brain, this area is the visual cortex that is located at the back of the brain (occipital lobe). The neural activity within the visual cortex would be related to consciously seen and imagined imagery, and the detection of this activity would be the basis for the external imaging of inner imagery. Obviously, external non-invasive and painless methods would be preferred for the detection of the neural activity; additional peep holes in the skull are not a popular approach.

Localized neural activity in the brain can be externally mapped by the so-called functional magnetic resonance imaging (fMRI) technology. A fMRI scanner consists of a large magnet, a radio frequency (RF) pulse generator and a receiver. During fMRI scanning the test person is placed inside the cylindrical scanner. Strong magnetic field is used to align the spins of hydrogen nuclei (protons) in the brain. A short RF pulse is used to change the spins, and at that moment the protons absorb energy. When the pulse is turned off, the spins realign to the magnetic field and the protons emit the absorbed energy as a radio signal. The decay time of this radio signal depends of the local ratio of de-oxygenated/oxygenated blood, which is supposed to be related to the local neural activity. This is the so-called *BOLD effect* (Blood Oxygenation Level Dependent effect) that depends on the inhomogeneities in the magnetic field caused by oxygen level changes in the blood.

fMRI technology allows the detection of neural activity with the spatial resolution of few millimeters or even less and the temporal resolution of few seconds. This allows the computation of scan images of the brain and its activity, but these images as such do not display any mental content or inner imagery of the scanned test person.

The next step in the imaging of the mental content with fMRI technology would be the interpretation of the scan images. More or less repeatable scan image patterns appear, when the test person views repeatedly the same or similar images. Statistical and associative methods can be used to find correlations between the seen objects and the scan image patterns. Thereafter probable images of the imagined objects can be constructed by these correlations. The principle of the associative reconstruction of mental content is depicted in Fig. 18.1.

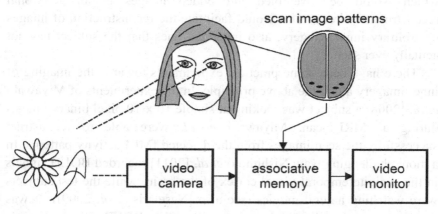

Fig. 18.1. The principle of the associative imaging of inner imagery. The scene that the test person sees is captured by a video camera and is stored in an associative memory. The test person's brain scan patterns are associated with the stored imagery and later on, when the test person imagines the same scenes, similar brain scan patterns are detected and these evoke associatively corresponding images at the associative memory. These images can be seen on a video monitor.

In Fig. 18.1 the subject and a video camera view the same object. The test person's visual cortex is scanned by the fMRI method and sequences of scan image patterns are achieved. The captured video images are associated with the brain scan patterns and are stored in the associative memory. The video images may be later on evoked associatively by similar brain scan patterns, when the test person is seeing or imagining similar objects. The evoked imagery can be seen on the video monitor.

The principle of the imaging of inner imagery is simple, but in practice there are some important points that should be noted. The eye has a small sharp vision area, the fovea, in the retina and this determines the instantaneous visual attention focus. The video camera should have a similar visual attention focus area, and the direction of the camera should track the gaze direction of the test person's eyes at all times. This would guarantee that the foveas of the eyes and the visual attention focus area of the camera would capture the same objects at each moment.

Furthermore, visual pre-processing should be used. Preferably, this preprocessing should extract similar visual features as the brain does. Consequently, the associative memory should contain visual features, which would be assembled into video images in an additional reconstruction circuit. This would facilitate the reconstruction of images of arbitrary inner imagery, also imagined ones that the subject has not actually ever seen.

There have been some practical experiments towards the imaging of inner imagery along the above principle. In the experiments of Miyawaki *et al.* [2008] a subject was looking at simple 10 x 10 pixel binary images during a fMRI scan. Miyawaki *et al.* were able to reconstruct successfully the seen images from the detected fMRI activity patterns. In a more challenging way, Nishimoto *et al.* [2011] recorded BOLD signals in the occipitotemporal visual cortex of the brain, while the test subjects were watching natural movies (see also [Naselaris *et al.* 2009]). It was found out that with some processing, the BOLD-signal patterns correlated well with the information in the movies. A Bayesian decoder was also constructed and used to translate the BOLD signal patterns into video images. According to Nishimoto *et al.*, remarkable reconstructions of the viewed movies were produced and the results would seem to verify that current fMRI technology is able to allow the decoding of dynamic brain activity.

The works of Miyawaki *et al.* and Nishimoto *et al.* would seem to confirm the connection between the physical brain processes and the mental content and also the hypothesis that the imagined and sensorily perceived images occupy the same neural area. These hypotheses are also basic tenets of this book and the basic idea behind the perception/response feedback loop model.

18.3. The Perception/Response Feedback Loop in the Imaging of Inner Imagery

The imaging of inner imagery is based on the correspondence between the visual sensory stimuli and the detected brain activity patterns that are caused by the perception of these sensory stimuli. This correspondence can be found out by subjecting the test person to well-defined visual stimuli, and recording these and the related brain activity patterns at the same time. In this way two tracks will be created, the brain activity track and the matching visual stimuli track. It is then assumed that the test person's imaginations will evoke similar brain activity patterns as the actual sensory perception of the same. In other words, the imaginations would correspond to the sensory stimuli. Based on this, the sensory stimuli that corresponds to certain brain activity patterns could be determined from the created brain activity track and the matching stimuli track. This sensory stimuli should then indicate and represent topic of the test subject's imaginations.

The practical realization of the above principle is not without problems. First, the sensory stimuli in the form of images and sounds should be matched with the corresponding brain activity patterns by recording these simultaneously. However, it may happen that the recorded stimulus images and sounds are not exactly those that have been consciously attended to by the test person. If they are not, the recorded stimuli track and the brain activity track will not correlate well. Therefore some mechanisms that would synchronize the focus of the stimulus image to the attention focus of the test person would be needed.

In practice, sensory stimuli are almost always somewhat different even when the same objects and events are perceived. Also the brain activity patterns differ. To overcome this problem, some kind of generalization and classification process would be required.

The use of the perception/response feedback loop principle will help to solve some of the above problems. Information processing with distributed feature signal representations and associative neurons would provide means to match the detected brain activity patterns with stimuli and to evoke the corresponding imagery in the form of video images. The soft detection function of the associative neurons may provide useful generalization and classification, if the features to be depicted by the signal representations are properly chosen. An ideal perception/response feedback loop for the imaging of inner imagery would be the one that closely matches the visual preprocessing of the brain.

The use of the perception/response feedback loop in the imaging of inner imagery is depicted in Fig. 18.2.

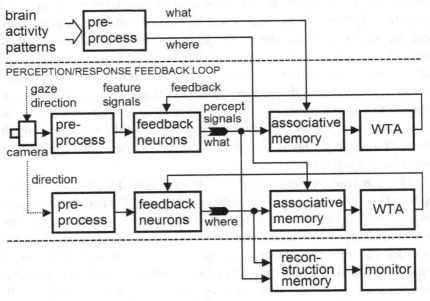

Fig. 18.2. The perception/response feedback loop in the imaging of inner imagery. Brain activity patterns are associated with corresponding visual percepts. Detected brain activity feature signals evoke corresponding visual "what" and "where" percept signals and these are transformed into viewable images in the reconstruction memory.

The system of Fig. 18.2 has two modes of operation, which are the learning mode and the imaging mode.

In the learning mode, the perception/response feedback loop associates detected brain activity patterns with corresponding visual features captured by the video camera. Direct video image is not well suited for this matching, therefore the image captured by the video camera is preprocessed into distributed feature signal arrays that depict elementary visual features at each position in the focus area of the visual scene. In this application the video camera should have a narrow field of acute vision that would match the fovea of the human eye. The gaze direction of the eyes of the test person would be detected, and the camera direction would be constantly adjusted to track and match the detected gaze direction. In this way the visual focus of the eyes and the camera would always coincide. The visual features captured by the camera would give the "what" information, and the gaze direction would give the "where" information. These representations are forwarded to the associative memories in the perception/response feedback loop. At the same time, the brain activity patterns (e.g. from fMRI) are preprocessed into distributed signal representations of "what" and "where". In the learning mode, these representations are then associated with the corresponding representations from the video camera in the associative memories.

In the inner imagery imaging mode, the detected brain activity patterns evoke corresponding visual "what" and "where" information at the associative memories. This information is transformed into viewable images by the reconstruction memory. In this mode, the camera and the related preprocesses are disabled and do not contribute to the overall operation.

A prerequisite for the correct operation of this system is the proper design and operation of the reconstruction memory. The reconstruction memory must be able to transform the distributed representations into viewable video. The proper functioning of this operation can be checked by letting the system reconstruct imagery from the camera image feature signals directly. If the direct reconstruction does not work properly, then the reconstruction of mental imagery will also fail. A personal computer may be used as the reconstruction memory and the display.

18.4. Detection of Inner Speech and Unvoiced Speech

The object of mind reading is not only the imaging of mental imagery. Another obvious target is the listening of the thought flow that appears in the form of the inner speech. The general principle of the imaging of the inner imagery can be applied to this task, too. The brain activity signals that could be used in this application would be those that control the production of speech. These signals should be found at the left side of the brain, at the so-called Broca area. However, the use of fMRI for this purpose would be challenging due to the limited temporal resolution of this method.

It has been found out that inner speech generates speech motor command signals even when the thoughts are not voiced aloud [MacKay 1992]. This phenomenon offers another possibility for the detection of inner speech. The speech motor commands go to the laryngeal area (neck) muscles that are responsible for the production of speech sounds. Activated muscles produce low-level electrical potentials that can be externally detected on the surface of the skin. This technique is known as *surface electromyography* (EMG).

Useful speech-related EMG signal patterns can be detected by the use of several properly placed electrodes at the neck and chin area. Strong signals are generated during voiced speech, but also almost spoken unvoiced speech and possibly the inner speech as such are able to generate detectable signals. It has been demonstrated that both voiced and unvoiced words can be determined from these signal patterns [Wand and Schultz 2009].

18.5. Silent Speech Detection with the Perception/Response Loop

The perception/response feedback loop can be utilized also for the listening of unvoiced speech with EMG signals. Inner speech may also be recognized by the same system, if the sensitivity of the EMG signal detection is sufficient and the preprocess is able to cope with the lower signal-to-noise ratio of the weaker EMG signals. The principle of a perception/response feedback loop based system for the listening of silent speech is depicted in Fig. 18.3.

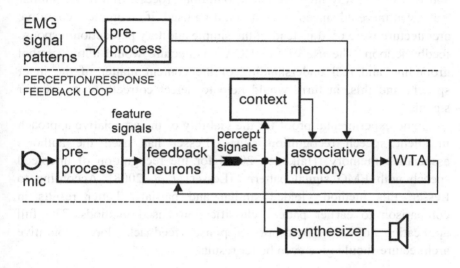

Fig. 18.3. The perception/response feedback loop system for the listening of silent speech.

The system in Fig. 18.3 has two modes of operation; the learning mode and the silent speech listening mode. In the learning mode, phoneme-related feature signals are extracted from the microphone signal and are associated with the preprocessed EMG signal patterns. Speech has temporal dimension, therefore the associative memory must be a serial one, accepting strings of signal patterns. Here autoassociation is used in addition to heteroassociation. This will allow the utilization of the context correlation between adjacent phonemes.

In the listening mode, the detected EMG signal patterns evoke strings of phoneme-related related feature signals. The most strongly evoked signals are selected by the Winner-Takes-All (WTA) threshold at the output of the associative memory. These signals are returned to the feedback neurons that produce percept signals. The percept signals are forwarded to the context memory and the speech synthesizer. The context memory contains a constantly updated string of the previous phoneme percept signals, which are used as a partial cue for the next phoneme. This will utilize the correlation between adjacent phonemes, and will in this way improve the recognition process. In a rather similar way even more advanced context could be used, if a complete cognitive architecture were used instead of the simple auditory perception/response feedback loop. The use of full HCA architecture would allow symbol grounding and real understanding of the content matter of the inner speech, and this, in turn, would help to detect correctly even weaker signals.

Some experimental proof for the viability of the associative approach in silent speech recognition exists. Lesser has used the author's associative memory [Haikonen 2007] for the recognition of unvoiced speech with EMG signal patterns [Lesser *et al.* 2008]. According to Lesser, this method yielded similar and partially better results in comparison to earlier pattern classification based methods. The full application of the perception/response feedback loop cognitive architecture should give even better results.

Chapter 19

A Comparison of Some Cognitive Architectures

19.1. Introduction

As illustrative examples of typical issues that relate to cognitive architectures, the following cognitive architectures are compared: Baars' Global Workspace architecture, Shanahan Global Workspace architecture and Haikonen Cognitive Architecture. Of these, the Baars model is the oldest, the Haikonen model is next and the Shanahan model is the most recent. The Baars model is well-known, and many other models have been derived from its general principles, such as the Baars-Franklin computational LIDA model [Baars and Franklin 2009], the CERA-CRANIUM model [Arrabales, Ledezma and Sanchis 2009] and later, the Shanahan [2010] model.

Also the Haikonen model has influenced some developments in this area, for instance the model of Kinouchi [2009]. The Kinouchi model is remarkable, because it is one of the few that explicitly consider the physical layer i.e. the executing artificial neural machinery and the phenomenal layer, "the logical layer".

The architectures of Baars, Haikonen and Shanahan share some common ideas and features, but there are also some decisive differences. All these architectures are claimed to offer some kind of explanation for the phenomenon of consciousness or at least for the difference between conscious and non-conscious operation.

In the following, the main principles and features of these cognitive architectures are summarized and compared with each other. Their ability to explain the real problem of consciousness, that is, the mechanism for the generation of the internal subjective appearance of neural activity in the form of qualia, is also compared. This is the most important and decisive point; a cognitive architecture that does not address this issue properly, is not a sufficient model for the mechanisms of a conscious mind.

19.2. Baars Global Workspace Architecture

19.2.1. Baars Concept of Consciousness

Baars sees the human consciousness as a biological adaptation that allows the brain to learn, interpret and interact with the world [Baars 1997]. Baars believes that the workings of the brain are distributed. There is no central command, instead the various networks of the brain are controlled by their own aims and contexts. All this has to be organized somehow and for that purpose there would be a network of neural assemblies that would display the contents of consciousness. Baars continues to propose that the cortical sensory projections areas would be best candidates for these networks. According to Baars the contents of consciousness include the perceptual world, inner speech and imagery, traces of immediate memory, feelings, autobiographical memories, intentions, expectations and so on. Baars proposes the concept of *focal consciousness*; having a mental event *that can be reported*, such as seeing an object or having a thought.

Baars proposes that consciousness has a function; it creates access to the information in various locations in the brain. Baars sees *consciousness as the publicity organ* of the brain. This organ would facilitate accessing, disseminating and exchanging of information. This organ would also exercise global coordination and control. Baars goes further: Consciousness prioritizes percepts, it is related to problem-solving by providing access to unconscious resources, it facilitates decision making, it optimizes the trade-off between organization and flexibility, it helps to recruit and control actions, it is needed for error detection and editing of action plans [Baars 1997, pp. 157–164].

19.2.2. Baars Model

The Baars Global Workspace Model [Baars 1988, 1997] is based on the idea of a network of neural assemblies that would display the contents of consciousness. This network would be a central working memory area, the so-called *Global Workspace*. The proposed workspace would act as a theater stage that would display and broadcast the contents of consciousness to the unconscious audience. This unconscious audience would consist of operators like memory systems, interpretation and recognition, action control, skills, etc. The Baars Global Workspace Model is depicted in Fig. 19.1.

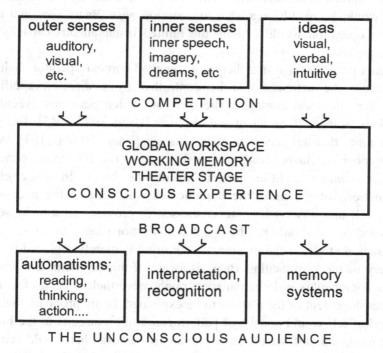

Fig. 19.1. Baars Global Workspace Model (compiled after Baars [1997]). The global workspace working memory is a kind of a theater stage that receives information from outer and inner senses and results from mental activity. The "spotlight of attention" selects the contents of the theater stage that is broadcast to the "unconscious audience". It is proposed that the broadcast part of the contents of the global workspace is also the contents of consciousness.

In the Baars model the contents of consciousness originate from outer senses (eyes, ears, touch, etc), inner senses (introspection of inner speech, imagery etc.) and ideas. Transmissions from these sources must compete against each other for the access to the global workspace. By Baars' definition, the contents of the global workspace memory constitute the conscious experience. The Baars model of conscious cognition is a *theater model*.

In the Baars model the global workspace working memory is a kind of a theater stage. Only a part of contents of the theater stage is consciously perceived at any time, namely the part that is "illuminated" by the "spotlight of attention". The rest of the contents of the theater stage would be readily available to consciousness. Baars sees that this kind of operation would explain the limited instantaneous capacity of consciousness.

Baars proposes also that there are so-called context operators behind the scene; the action would be controlled by a director, spotlight controller and local contexts. The director, which performs executive functions, is the self, an agent and observer [Baars 1997 p. 45]. The self is the entity that has access to consciousness [Baars 1997 p. 153]. With this proposition Baars comes perilously close to the discredited concept of *homunculus*, "a little human" inside the brain. In some early explanations of the mind the homunculus was seen as the necessary acting self that observed whatever the senses provided to it to be seen, heard and felt. It should be obvious that this is not a logically satisfactory explanation at all, as the mystery of the mind is merely replaced by the mystery of the homunculus. Baars is aware of this problem, and argues that a homunculus-style explanation might nevertheless be useful, if it reduces the extent of the matters to be explained. Baars proposes that the observing self would consist of pattern recognition circuits in the brain. In general, the self would be a mental framework that would remain stable across the course of life [Baars 1997, pp. 142–153]. The self would be able to will and plan, prioritize, make decisions and solve problems. According to Baars, it is consciousness that facilitates these functions. It is not immediately obvious that this reasoning is free of any *circularity*, the act of explaining things by themselves.

The Global Workspace Model is also a *Blackboard Model*. Baars [Shanahan and Baars 2005] states that the global workspace architecture was inspired by and is derived from earlier Artificial Intelligence blackboard systems like those presented by Hayes-Roth [1985] and Nii [1986]. In Artificial Intelligence the Blackboard Model is a method of computation with a group of specialist operators. A problem is "written" on the "blackboard", a common working memory, where it is available for all the specialist operators. Each operator reads the blackboard and tries to update the blackboard presentation whenever it can provide a step towards the final solution. However, at each moment only the specialist that has the most relevant update, may write on the blackboard; therefore the specialists must compete against each other for the access to the blackboard.

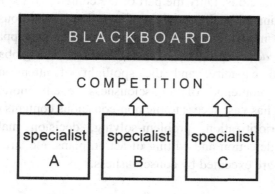

Fig. 19.2. The Blackboard system. The blackboard displays the problem at hand and also the intermediate steps towards a solution. The specialist that has the best contribution towards the solution at each moment may write its contribution on the blackboard for everybody to see.

The blackboard model is depicted in Fig. 19.2. At each moment any of the specialist operators A, B, C may write on the blackboard if its contribution is more relevant than the contributions of the other specialist operators. (The number of specialist operators is not limited to three in actual applications.) The blackboard "displays" or "broadcasts" continuously its contents to every specialist operator. It can be seen that the blackboard method has an important restriction: On the blackboard the information must be presented in a common code or language (lingua franca) that every specialist operator can understand and use. However, in their internal processes the specialist operators may use whatever codes are suitable.

19.2.3. What Is Explained by the Baars Model

According to Baars, consciousness is created by the broadcasting of the information selected by the "spotlight of attention" that "illuminates" some of the contents of the theater stage, also known as the global workspace working memory.

The difference between conscious and non-conscious activities is explained as follows: Most of the mental activity and contents of the brain is non-conscious. Only the part of the contents of the theater stage, which is illuminated by the spotlight of attention is conscious.

The Baars model proposes an explanation to the apparent *limited capacity, vast access* property of consciousness. The global workspace has a limited capacity, and the spotlight of attention limits the instantaneous contents of consciousness even more. However, consciousness has vast access to the non-conscious contents of the brain.

Percept prioritization, problem-solving, decision making, action control, error detection and editing of action plans, etc. are explained as functions that are executed by consciousness.

19.2.4. What Is Not Explained by the Baars Model

The Baars model is a general one, more like a metaphor, which does not give details of the actual workings of the underlying machinery. In the brain the machinery would consist of neurons and neuron groups, but the Baars model does not define these, nor propose any detailed method of interaction between these.

The Baars' Global Workspace model does not explain, why and how the neural activity could appear internally in the form of qualia. Thus the Baars Model does not address the fundamental issue of consciousness at all and therefore is not a sufficient model for conscious agents.

Summary of Baars Global Workspace Architecture

- The Baars Global Workspace Architecture is a theater model.
- The Baars Global Workspace Architecture is derived from earlier Artificial Intelligence Blackboard models.
- The Baars Global Workspace Architecture consists of a number of specialist modules and a common working memory area, the so called Global Workspace.
- The Global Workspace broadcasts its contents to all specialist modules.
- Each specialist module competes for the right to transmit its contribution to the Global Workspace; only the most relevant contribution is accepted at a time.
- The Global Workspace model explains the difference between conscious/non-conscious activity.
- The Global Workspace model explains the serial narrow bandwidth nature of the stream of consciousness; specialist modules operate subconsciously in parallel way, while the stream of consciousness mediated by the global workspace is serial.
- Baars' consciousness has a function; it facilitates accessing, disseminating and exchanging information and exercises global coordination and control.

19.3. Shanahan Global Workspace Architecture

19.3.1. Shanahan Concept of Consciousness

In his nice book *Embodiment and the Inner Life* Murray Shanahan has presented an alternative Global Workspace model, which is inspired by the Baars model [Shanahan 2010]. Rather than defining consciousness, Shanahan tries to approach the phenomenon by inspecting the difference between conscious and unconscious processes. Shanahan sees that unconscious actions are kinds of automatic actions that are executed without conscious attention. On the other hand, conscious actions are characterized by introspective reportability, enhanced flexibility in novel situations, the mental ability to execute problem-solving steps and the ability to lay down memories.

Shanahan sees that in the brain there are many simultaneously ongoing activities. According to Shanahan, an important hallmark of the conscious condition is *unity*, the integration of these different brain activities so that they can influence each other. Shanahan proposes that in the conscious condition, the brain's multitude of processes act as an integrated whole. Shanahan sees that in the brain there is a mechanism that facilitates this integration, which leads to the conscious condition; this mechanism is the global workspace. Shanahan's global workspace is, however, different from that proposed by Baars.

19.3.2. Shanahan Model

The basic components of the Shanahan and Baars models are the same; a number of specialist modules and a global workspace. The specialist modules would try to broadcast their information to the global workspace. However, Shanahan recognizes that the theater models of mind border on the concept of the discredited homunculus. Therefore he rejects the theater models of consciousness and, accordingly, the role of the global workspace as a theater stage. Consequently, the Shanahan architecture is not a theater model.

In the Baars model the global workspace is a working memory that acts as a theater stage, where the conscious contents of the mind is displayed to the audience of the non-conscious modules. Shanahan rejects this and proposes that instead of a theater stage and a common working memory area, the global workspace should be thought as a communications infrastructure. This infrastructure would connect the various autonomous specialist units to each other, and in this way would facilitate the brain activity integration that is a supposed prerequisite for consciousness [Shanahan 2010 p. 111].

However, in the Shanahan model the global workspace is more than a mere communications infrastructure, as it executes also other functions. Shanahan proposes [Shanahan 2010 p. 148] that besides being the locus of broadcast, the global workspace is also the arena for competition between rival connections of modules. The essential elements of the Shanahan model are depicted in Fig. 19.3.

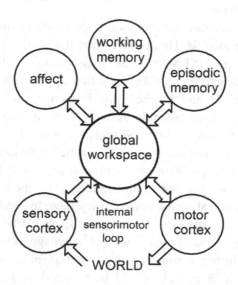

Fig. 19.3. The Shanahan model (compiled after Shanahan [2010]). The Shanahan global workspace is a communications infrastructure that connects the various modules with each other. The modules must compete for the access to the global workspace. An internal sensorimotor loop allows the mental simulation of actions. It is proposed that the information that goes through the global workspace constitutes the contents of consciousness.

Figure 19.3 depicts the main modules and interconnections of the Shanahan model. In this model the global workspace infrastructure connects the modules of affect, working memory, episodic memory, sensory cortex and motor cortex to each other. Shanahan assumes that this division of functions would roughly conform to the anatomical divisions in the brain. The sensory cortex and the motor cortex form a functional loop, which is closed by the external world; the actions executed by the motor cortex can be perceived by the senses and the sensory cortex.

Hesslow [2002] has proposed that thinking is simulated action and perception, which is enabled by an internal sensorimotor feedback loop. Shanahan has augmented his model by the inclusion of an internal sensorimotor loop in the style of Hesslow. This loop would allow the internal perception of planned and imagined actions without the need to actually execute these.

Shanahan is aware of the common code problem that is inherent in global workspace models. He proposes that the need for a lingua franca [Shanahan 2010 p. 118] within the framework of global workspace communication structure could be remedied easily. If a module A is to influence modules B and C, then this influence can be mediated by different signals; i.e. the module A would use dedicated codes when transmitting to different modules. Shanahan states: "The signals going to B and C from A do not have to be the same". Obviously this method bypasses the original problem of common code, but at the same time it creates further problems and complications. The modules cannot now broadcast one universal signal pattern, they have to send different signal patterns to different modules. Consequently, each transmitting module would now have to master a number of languages instead of a single universal one. During communication the transmitting module would have to know, which module would be targeted, and it would have to generate a specific signal pattern for that module. However, associative communication could also be utilized in Shanahan's model. This would remedy the common code problem.

19.3.3. What Is Explained by the Shanahan Model

Shanahan proposes that the global workspace communication infrastructure explains the conscious/unconscious distinction in a way that is compatible with the ideas of Dennett and Tononi. According to Dennett [2001] the difference between the conscious and unconscious processing would arise from the difference between local and fully global influence; a conscious process would recruit global resources. Tononi's [Tononi *et al.* 1998] information integration theory states that a conscious process would involve global interactive neural activity.

Along these lines, Shanahan proposes that the information mediated by the global workspace communication infrastructure constitutes the contents of consciousness, while the specialist units operate unconsciously. The global workspace communication infrastructure would allow an integrated response, and it would enable learning and episodic memory making, etc. by allowing the various modules to cooperate on the same topic. Shanahan states: "Perfect integration occurs when the being as a whole is brought to bear on the ongoing situation" [Shanahan 2010 p. 112]. The limited bandwidth of the global workspace communications infrastructure would allow only one coalition of processes at a time. This limitation would direct the focus of attention onto a single, unified object and lead to the forming of a single, unified thought. In this way the stream of serial consciousness would arise from the parallel processes of the brain.

Unfortunately, the allowing of only one coalition at a time comes at a cost. In addition to the ongoing conscious experience, conscious humans are able to process additional semiconscious or sub-conscious tasks at the same time. There may be some sub-consciously processed tasks within the various modules, but there are also tasks that call for limited coalitions of modules. For instance, the reading of a text calls for sensorimotor coalitions that control the visual attention and the movement of the eyes. However, this action is not consciously perceived, unless the reader specially focuses attention on this. Yet, at the same time, other coalitions must process the meaning of the text; the results from these coalitions would be consciously perceived. In Shanahan's model this could not happen; simultaneous, but separate coalitions are forbidden.

The internal sensorimotor loop explains the ability to imagine motor actions as if they were actually executed. This function would allow the planning and selection between suitable actions.

19.3.4. What Is Not Explained by the Shanahan Model

The Shanahan Model does not explain, why and how the supposedly conscious neural activity in the global workspace communications structure could appear internally in the form of qualia. Thus, the Shanahan Model does not address the fundamental issue of consciousness and therefore is not a sufficient model for conscious agents.

Summary of The Shanahan Global Workspace Architecture

- The Shanahan Global Workspace Architecture is not a theater model.
- The Shanahan Global Workspace Architecture consist of a number of specialist modules that are connected to each other via a communications infrastructure, the Shanahan global workspace.
- The communications infrastructure is a narrow bandwidth connective core.
- The specialist modules must compete for the access to the connective core.
- Specialist modules operate non-consciously. Consciousness occurs when the communications infrastructure mediates an integrated response by bringing the whole system to bear on the ongoing situation.
- The information mediated by the global workspace communication infrastructure constitutes the contents of consciousness.
- The Shanahan model includes an internal sensorimotor loop that allows the imagination of motor responses.

19.4. Haikonen Cognitive Architecture

19.4.1. Haikonen Concept of Consciousness

The Haikonen cognitive architecture is based on the following hypotheses about consciousness:

- Consciousness is the reportable presence of self-explanatory internal appearances (qualia) of the direct and externalized sensory percepts of the environment, the body and the virtual percepts of mental content. Higher cognition is based on the symbolic use of these.

- Reportable percepts constitute the contents of consciousness; without percepts there is no consciousness.
- A conscious mind has the flow of sensory and imagined percepts and possibly a meaningful flow of symbolic percepts such as a natural language inner speech.
- A conscious mind has the ability to report its contents to itself and outside in various ways and possibly by using a natural language.
- Consciousness is an inner appearance of information, it is not an agent.

Higher symbolic cognition and the mastery of a natural language are not seen as necessary prerequisites for consciousness. However, it is recognized that a plausible model of cognition and consciousness should be able to explain also these.

19.4.2. Haikonen Model

The Haikonen Cognitive Architecture is primarily a blueprint for artificial robot brains. As such, it details both the overall architecture, the information processing method and the basic building blocks. It is inspired by the human brain and cognition, but does not seek to model the brain accurately.

The Haikonen Cognitive Architecture tries to address the qualia-aspect of consciousness; it tries to create inner appearances of the perceived world via direct sensory perception and also inner appearances of selected mental content via the feedback principle. Technical details of the Haikonen model are given in [Haikonen 2007]. More general details as well as cognitive and philosophical background can be found also in [Haikonen 2003].

The Haikonen Cognitive Architecture is designed to fulfill the requirements of perception and sensorimotor integration, the flow of sensory and imagined percepts, emotional effects and a natural language. In robotic implementations it has means to perceive its environment and body, introspect and report its mental content, understand situations and their requirements, plan and control motor actions and exercise judgment with emotional values.

The Haikonen architecture is a parallel and distributed one, without any global workspace or "a theater stage". There is no common code either. The executive and attention functions are distributed. The architecture consists of a number of modules for sensory and motor modalities. These modules are rather similar to each other, and are based on the principle of the perception/response feedback loop. Each sensory modality has its own perception/response feedback loops that can handle a large number of parallel signals.

19.4.3. What Is Explained by the Haikonen Model

Qualia: The Haikonen model is founded on the proposition that consciousness is based on perception. Direct perception, as opposed to indirect perception via symbols, is assumed to facilitate some kind of qualia. The perceptual information is carried by neural signals, but the system is not able to perceive the signals as such; neural signals are only transparent carriers of the information about the sensed features. The carried pieces of information appear as the machine qualia. These qualia are not necessarily similar to the human qualia, with the possible exception of amodal qualia.

The difference between conscious and non-conscious activities is explained as follows: The associative cognitive system has continuously ongoing activities; some of these are related to sensory perception and appear as percepts. Some activities are related to forking chains of associations "deep" within the system. Reportable percepts are conscious while the association chains within the system are not. The products of the non-conscious activities may become consciously perceived when they are transformed via the feedback loops into virtual sensory percepts.

The reportability aspect of consciousness is explained as follows: Percepts are broadcast to every other modality; the receiving modality determines which broadcasts are accepted and attended to. When percepts of a certain event are accepted by many modalities, multiple associative connections may be formed, and consequently memories of that event will be created. This allows the reporting of that event in terms of the receiving modalities. Thus attended percepts will be remembered for a while and can be reported; this is one criterion for conscious percepts.

The limited serial nature of consciousness is explained as follows: Basically, the associative system is a parallel network that is able to process many things simultaneously. However, in the conscious mode the various modules are focusing their attention on the same thing. Only one thing (or few things depending on the actual realization of the system) can be globally attended to at a time, therefore the flow of consciously attended percepts will be temporally serial. Also practical reasons limit parallel operation; it is not possible to speak two words at the same time, therefore there is only one stream of inner speech.

Cognitive functions: The Haikonen model proposes means for the artificial realization of a number of cognitive functions, such as perception, learning, memory making, introspection, language and inner speech, imagination, emotions.

Summary of the Haikonen Cognitive Architecture

- The HCA is not a theater model or a global workspace model.
- The HCA is a sub-symbolic/symbolic neural system.
- The HCA utilizes self-explanatory sensory information for the grounding of the meanings of the symbols.
- The HCA consists of a number of specialist modules; perception/response feedback loop units, that are connected to each other via wide bandwidth signal lines.
- The specialist modules broadcast their percepts.
- The specialist modules decide when to accept broadcasts and from whom.
- Specialist modules operate non-consciously.
- When modules focus on the same object and situation ("information integration"), allowing cross-association and evocation, the reportability aspect of consciousness is realized.
- The HCA incorporates inner speech.
- The HCA incorporates system reactions that facilitate pleasure and pain according to the SRTE model.

19.5. Baars, Shanahan and Haikonen Architectures Compared

The Baars, Shanahan and Haikonen architectures utilize a number of non-conscious autonomous specialist modules that interact with each other in one way or another. In the Baars model this interaction takes place via the global workspace memory, the "theater stage", which broadcasts the attended contents to the specialist modules. In the Shanahan model the interaction takes place via the global workspace communications structure and in the Haikonen model the modules communicate directly with each other.

Figure 19.4. depicts the general principle of the internal communication between the autonomous modules and the global workspace working memory in the Baars model. The global workspace working memory broadcasts the attended part of its contents to each autonomous module. The modules must compete against each other in order to be able to post their information to the global workspace working memory and eventually via this route to the other modules. The number of the autonomous modules is not limited to the four modules A, B, C, and D shown in this depiction. Further modules would be connected to the global workspace in a similar way.

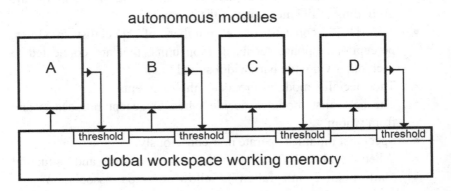

Fig. 19.4. Internal communication in the Baars Model. The transmissions from the autonomous modules A, B, C, D compete against each other for the access to the global workspace working memory. The global workspace broadcasts its contents to every autonomous module. Common code is required.

Figure 19.5 depicts the general principle of the internal communication between the autonomous modules and the global workspace communications structure in the Shanahan model.

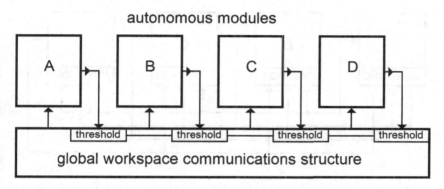

Fig. 19.5. Internal communication in the Shanahan Model. Transmissions from the modules compete against each other for the access to the global workspace communications structure. The global workspace broadcasts its contents to every module.

The modules of the Shanahan model must compete against each other for the right to input information the global workspace communications structure. This information is received by the other modules from the global workspace communications structure. The global workspace communications structure has limited bandwidth and allows the formation of one coalition of processes at a time, the system cannot support several simultaneous coalitions.

Figure 19.6 depicts the general principle of the internal communication in the Haikonen model. In the Haikonen model autonomous modules broadcast their information directly to each other. A receiving module determines, which (if any) of the simultaneously available transmissions is accepted at each moment. A module may also solicit a certain response by transmitting suitable cues that may evoke a strengthened response and transmission in the solicited module. Common code is not required.

In the Haikonen model, simultaneous separate coalitions of modules are allowed and these ad hoc cooperative coalitions may form as the need arises. This property facilitates the execution of several tasks at the same time. However, only one task of these may be executed in a way that displays the hallmarks of consciousness.

autonomous modules

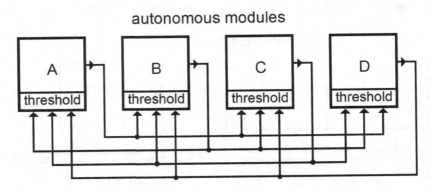

Fig. 19.6. Internal communication in the Haikonen Model. Each autonomous module broadcasts its output to every other autonomous module. Each module has its own input threshold that determines which (if any) of the simultaneously available broadcasts is accepted at each moment. Arbitrary "ad hoc" coalitions of modules are possible.

In the Baars and Shanahan models the global workspace can be understood as the site of conscious mental content. In the Haikonen model there is no box that could be labeled as "consciousness"; every module operates in the same way whether the overall action is conscious or not. The hallmarks of consciousness would arise from the use of self-explanatory information and reportability, allowed by short term memory and the cooperation between modules.

The Baars, Shanahan and Haikonen models propose that the focus of consciousness is determined by global attention. However, the Haikonen model does not utilize a global workspace, which is seen as redundant, because the modules can communicate and compete directly with each other. There are also other important differences. Baars and Shanahan models do not address the real problem of consciousness; how the inner subjective experience arises from their processes. They do not really explain qualia, the inner appearance aspect of consciousness. They do not specify how inner speech could arise, either. The Haikonen model is specific on these issues.

Chapter 20

Testing Artificial Consciousness

20.1. Requirements for Consciousness Tests

Basically, there is only one fundamental requirement for real human-like phenomenal consciousness, and that is the requirement of the presence of qualia-based subjective inner experience, "the internal appearances" of neural or electronic activity. Unfortunately this hallmark of consciousness is available to and is observable only by the subject itself, unless some ingenious methods are invented. Therefore, currently the possible consciousness of a robot can be determined only by indirect ways.

The presence of consciousness in human subjects can be tested quite simply, because we know apriori that the subject is a conscious being. Therefore the subject is either unconscious or conscious (partly or fully), and we only have to find out, in which state the subject is at the moment of testing. Thus, for instance, when the subject is apparently asleep or may have lost consciousness due to anesthesia or some medical condition, we may try to arouse the subject and test the following.

- Does the subject have a functioning perception process; does the subject feel pain, does the subject see and/or hear, does the subject respond to stimuli?
- Is the subject able to report anything, verbally or by using bodily signs?
- Are the reports sensible or only reflexes?
- What is the extent of the subject's situation awareness? Does the subject know his/her name, where he/she is, what day it is, etc.

These tests usually work, and the subject is more or less conscious if responses are found. However, a subject may fail in these tests and yet be conscious in limited ways. A sleeping person may not respond to these tests, yet the person may be aware of an ongoing dream that she/he may be experiencing. An apparently unconscious patient may be in a "locked-in state", so that she/he cannot respond to stimuli and report pain even though she/he is actually experiencing those.

The detection of artificially generated consciousness is more difficult, as clearly explained e.g. by Musser [2016]. In this case we cannot begin with the assumption that the artifact has a mind with non-conscious and conscious states. Instead, this is exactly what we seek to find out. A cognitive robot (such as the author's XCR-1 robot) may have various sensory modalities and perception processes, and it may generate verbal reports showing that the sensors are acquiring valid information. Yet, without further investigations we may not know, how the sensory percepts appear internally; is there any "subjective experience" at all. Therefore the above tests may not be reliable. If a robot fails to respond to these tests, then it is more credible than not that the robot is not a conscious agent. On the other hand, if the robot passes these tests (like the XCR-1 robot), it is still not necessarily a conscious agent, because these tests can be passed in many mechanical ways, without any real qualia-based consciousness. Thus, instead of direct tests like the ones above, we have to use indirect methods to determine the presence and scope of consciousness in a cognitive robot.

There are some common sense preconditions for the existence of consciousness in any agent. Therefore the evaluation of a robot's consciousness should begin with the study of the robot's cognitive machinery, the robot's brain. This study should determine, whether the machinery could support the creation and presence of phenomenal qualia-based mental content. Some general criteria exist that do not call for the full understanding of the nature of qualia.

General awareness calls for the existence of perception process with a number of sensory modalities. The perception processes must be direct and sub-symbolic. Percepts must not be based on digitalized symbolic representations, like binary numbers, because these would not be self-explanatory. There is also the requirement of mental content; an agent cannot be aware of its mental content if there is none. A conscious agent must also be able to report its mental content to itself; it must be able to introspect its own thoughts. Perception and introspection call for means of selection, because all information cannot be treated at the same time. Therefore the function of attention must be included. Consciously attended events can be remembered for a while, thus memory is also needed.

In addition, a conscious agent should be able to generate responses to the on-going situations and also be able to report its actions and intentions in one way or another to its masters. However, this is a secondary requirement arising from practical causes, as an agent or robot without these capacities would not have much practical value. Nevertheless, the ability to report would help in the evaluation of the agent's possible consciousness.

Therefore the cognitive machinery of a potentially conscious agent should support the following general requirements for consciousness:

- Perception, direct, self-explanatory
- Mental content
- Introspection
- Attention
- Memory and retrospection
- Responses and reports

The presence of these functions in a cognitive machinery should be rather easy to see. If these functions are not realized within the cognitive machinery of a robot, most probably the robot will not be able support any kind of qualia and be conscious in the human way. However, the presence of these functions does not necessarily guarantee the presence of any consciousness with qualia.

A robot with the above functions, but without qualia and inner experience might behave in a way that would create the external appearance and impression of a conscious agent. Some additional cognitive functions might even lead to the impression of an intelligent agent with free will. This kind of a robot could be said to be *functionally conscious*. However, functional consciousness is not real consciousness, instead it should be seen as some kind of a behavioral simulation of the external appearances of the real thing. Here "functional" means "as if".

20.2. Tests for Consciousness

20.2.1. The Turing Test

The Turing test [Turing 1950] is sometimes proposed as a test for machine consciousness, as such or with modifications. In the fifties Alan Turing was considering the question "Can computers think", and if they do, how could we know. In those days computers were readily likened to the brain; they were electronic brains executing mental actions that earlier could only be executed by the thinking human mind. For instance, the conditional IF – THEN program command was naively seen as the computer's ability to make decisions.

Turing proposed that a computer could be said to think, if it were able to converse with a test person in such a way that the person would actually believe that the other party is a human. It should be obvious that this test is badly flawed; to make believe is not the same as to make true, see Fig. 20.1. Besides, this kind of fooling tells nothing about the real requirement of consciousness, the presence of qualia-based subjective inner experience.

Fig. 20.1. The problem with the Turing test. Who is actually being tested, and for what?

The so-called *Total Turing Test* [Harnad 1992] is supposed to remedy the obvious flaw in the original Turing test. In order to pass the Total Turing Test the artifact is required to perform exactly as a human being in every empirically testable cognitive challenge; indistinguishable to any judge, and for a lifetime. The totality of this test is supposed to prove that the artifact is indeed a thinking and aware being, cognitively similar to humans. However, the Total Turing Test is based on logical fallacy; it is logically true that an artifact that is cognitively equivalent to humans will pass this test, but it is still logically possible that this test is passed by artifacts that are not conscious and cognitively similar to humans. The Total Turing Test is based on external appearances, and as such it does not indicate, which group the tested artifact would belong to. It is of no use to argue that the Total Turing Test would be so exhaustive that no other than those artifacts that are cognitively similar to humans could pass the test, *without also proving first* that the passing of this test by any cognitively non-similar, non-conscious artifact were impossible.

A test that does not work in the first place will not work any better if the scope of the tested property is extended. There is nothing in the Turing Test or the Total Turing Test that could directly test the presence of any qualia-based inner experience.

20.2.2. Picture Understanding Test

According to the Information Integration Theory of Consciousness, proposed by Tononi [2004, 2008], conscious information is integrated and unified. This unification arises from a multitude of interactions between the different parts of the brain. If the interactions cease, as during anesthesia or deep sleep, consciousness disappears. According to Tononi, information integration is necessary for consciousness. Also, a subject needs a large repertoire of actively connected information in order to be conscious. Based on that, Koch and Tononi [2011a, 2011b] have proposed that the presence and amount of information integration can be used to test consciousness in machines by letting them to try to understand pictures.

Koch and Tononi propose an example: Let us suppose that there were two test pictures. One picture depicts a computer display with a keyboard in front of it, and the other picture depicts the same computer display with a flower pot in front of it. According to Koch and Tononi, a non-conscious computer with machine vision would not realize that there is something wrong with the second picture, because the computer would not have and would not be able to evoke the necessary background information about the relationships of the items in the picture. Therefore the computer would not really understand what it sees. On the other hand, we, conscious humans, would immediately see and recognize the untypical situation due to our vast amount of background information and our capacity to integrate it meaningfully.

It should be obvious that this test is the Turing test in disguise. To pass this test the machine should exhibit similar cognitive abilities in picture understanding as humans. However, there nothing in this test that could prove that the machine was actually conscious; this test addresses only certain issues of computer vision algorithms and the integration of any available background information. The example test could be easily passed by artificial neural classification, by storing a large number of pictures of computer displays with keyboards. Thereafter a picture of a computer with a flower pot would surely stand out as an exception. No consciousness would be involved here.

As such, this test does not have much to do with consciousness, because the essential issues of qualia and inner experience are ignored.

20.2.3. The Cross-Examination Test

It seems that currently there is only one way of finding out the presence of qualia-based inner experience in a robot; we must ask and cross-examine the robot itself. Unfortunately this method would exclude those robots and machines that do not use and understand a natural language; this is a similar problem to that, when we try to find out whether animals have any inner experiences.

If the robot is able to report that it has the flow of inner imagery and inner speech, and we know that this report is not a pre-programmed one and the internal architecture of the robot's brain could be able to support these, then we should determine that the robot is actually able to introspect its mental content and report it; the robot is conscious. If the robot is also able to report that it has inner speech, which appears as a kind of heard speech, then we should determine that the robot has some kinds of internal appearances and qualia. The cross-examination test may focus on the following issues:

- Does the robot have a flow of mental content that is about something?
- Is the robot able to report its mental content (percepts, thoughts, inner speech etc.) to itself (and others) and does it recognize the ownership of the same?
- Is the robot able to report some qualia?
- Does the robot remember its immediate past?
- The "hammer test" of phenomenal awareness: Does the robot feel pain? In which way?

One may wonder, if the cross-examination test were just another variant of the Turing test. In the Turing test the actual realization of the tested machinery is not considered. The cross-examination test, however, is meaningful only for architectures that are known to fulfill the basic requirements for potentially conscious machines.

20.2.4. The ACT Test

Schneider and Turner [2017, 2019] have proposed a test for machine consciousness, called the AI Consciousness Test (ACT). This test is applicable to potentially conscious machines and robots that can communicate by a natural language. In this test the machine is subjected to increasingly demanding conversation about concepts, scenarios and consciousness-related internal experiences. Ultimately, the machine's ability to understand and discuss philosophical questions about consciousness would be tested. A very positive sign of the presence of consciousness would be the machine's ability to invent and use consciousness-related concepts on its own, without the help of ideas and inputs from humans.

20.3. Tests for Self-Consciousness

20.3.1. Testing Self-Consciousness

First of all, the testing of self-consciousness is meaningful only if the test subject (robot) is known to be phenomenally conscious, and thus be able to utilize self-explanatory information for the grounding of meanings. If the test subject is not phenomenally conscious, then the results of self-consciousness tests indicate only the presence or absence of certain mechanistic responses.

The testing of self-consciousness of a robot should start with the verification of the presence of the equivalent of the somatosensory system and memory systems with the functions of introspection and retrospection. This can be verified by the inspection of the cognitive architecture of the robot. If these are implemented in the robot, then the presence and extent of a body image and a mental self-image should be determined.

Sometimes the block diagram of a cognitive system may contain "boxes" with the labels "body image" and "self-image", or even "consciousness". These boxes should not be taken at the face value. The presence of the body image and mental self-image should be tested by functional means, for instance, by checking the robot's ability to refer to its body and self-related mental content.

20.3.2. The Mirror Test

The so called *mirror test* is sometimes used to test the self-consciousness of babies and also animals [Gallup 1970]. In this test, the test subject is placed in front of a mirror and the subject's behavior is observed. The test is passed, if the subject recognizes itself in the mirror. Many animals and very young babies seem to take their own image in the mirror as another being, and may try interact with these and look behind the mirror. In these cases the test is clearly not passed. Sometimes, however, it may be difficult to determine directly, whether the subject recognizes itself in the mirror, and therefore variations of this test may be used.

The *rouge test* is one variation of the mirror test. It can be used as soon as the test subject has been familiarized with mirrors and has seen its own mirror image repeatedly. In this test a red dot is secretly placed on the forehead of the subject to be tested. Next, a mirror is placed in front of the subject. The rouge test is passed, if the subject realizes that the red dot on the mirror image depicts the red dot on the subject's face. Human babies of the age of 18 months or over usually pass the mirror test. This is also the age, when babies usually become socially self-aware [Lewis *et al.* 1989].

Some animals including chimpanzees, elephants and possibly pigeons can also pass the test. In the strict sense, the mirror test only indicates that in some cases the test subject is able to recognize itself in the mirror image. This is a cognitive task that may fail even when the subject actually has some kind of self-concept. The mirror test has been applied to robots by, for example, Takeno *et al.* [2005].

20.3.3. The Name Test

The *name test* is based on the assumption that if a subject has a self-concept, then this may be associated with a name. Little children learn their names early, and may use those instead of "I" to refer to themselves in their speech.

Many animals seem to be able to learn their given names, but it is not clear, whether they associate this name with themselves or with an event. A cat may come when called by name, but this may happen, because in the cat's mind the name is actually associated with food or petting, not with the cat itself. Animals without the ability to speak cannot use their names in overt speech to refer to themselves, and thus we will not know, what kinds of associations have taken place. Therefore, the one-way passing of the name test is not necessarily a reliable indication of self-consciousness.

The name test can be used with robots, too, if the auditory modality allows the recognition of at least some spoken words. If an advanced robot is able to use a natural language, then it should be possible to observe, if the robot has learned its name and uses it to refer to itself. But then, it would also be possible to ask the robot about its concepts of self, and in this way the issue of the robot's self-consciousness could be resolved.

20.3.4. The Ownership Test

A self-conscious subject should posses the concept of the ownership of the body, thoughts, memories, decisions, speech, acts and possible acquired possessions. One indication of the presence of the concept of ownership might be the subject's ability to state that "this is mine" or "I did this" meaningfully, not as a recorded or rote-learned message.

If the subject is not able to communicate with a natural language, then the presence of the concept of ownership must be determined by indirect means. Does the subject recognize that a certain act has been executed by the subject itself? The determination of this may not necessarily be impossible. Consider, for instance, a dog that has destroyed few household items just for fun, while left home alone. When confronted by the master, the dog may readily show the apparent signs of guilt and shame; the dog apparently knows and remembers, who has done all this.

20.3.5. The Cross-Examination Test

The cross-examination test may also be used to determine the presence and extent of the self-consciousness in a robot that is able to communicate verbally. For instance, the following questions may be considered:

- Is the robot able to make the difference between the environment and the robot self?
- Is the robot aware of its own existence? In which way?
- Does the robot relate its present situation to the personal history and to the expected future?
- Self-description; body image and mental self-image. How does the robot perceive itself?
- Ownership; is the robot able to declare any ownership?
- Existence; is the robot afraid of its destruction?

20.4. Axioms and Assessments for Machine Consciousness

20.4.1. Aleksander's Axioms

Aleksander and Dunmall [2003] propose five axioms that would define a minimal set of material preconditions for a conscious biological or non-biological agent. These axioms are:

1. Depiction. The agent must have internal perceptual states that depict external world and the agent itself.
2. Imagination. The agent must have internal states that depict recalled or imagined depictions of the style of the axiom 1.
3. Attention. The agent must be able to choose which parts of the world or the agent itself is internally depicted at a given time.
4. Planning. The agent must be able to imagine future action and select among imagined alternatives.
5. Emotion. The agent must have affective states that evaluate planned actions and determine the selection of the action to be executed.

Obviously humans usually satisfy these requirements and are also conscious. The first requirement, depiction, would seem to be the most important one, because without any internal perceptive depiction the requirements 2–5 could not be fulfilled, and no cognition could take place.

However, humans that have limited faculties of imagination, attention, planning and emotion, would still be counted as conscious persons albeit not very bright ones. On the other hand, a cognitive machine might seem to posses these capacities, and yet be without any real consciousness; it would all depend on the nature and style of the internal depiction; is it qualia-based or not.

Aleksander's axioms are a useful checklist of necessary preconditions for machine consciousness, but the fulfillment of these preconditions in a system does not necessarily guarantee the presence of consciousness in that system.

20.4.2. The ConsScale

Arrabales, Ledezma and Sanchis [2010] propose that consciousness is a rather continuous phenomenon, starting from a very weak and dim minimal consciousness, and extending to a possible super-conscious state. Accordingly, Arrabales *et al.* have devised a biologically inspired scale for the assessment of the level of the functional consciousness in machine consciousness implementations. This scale, *ConsScale*, is a partially ordered set of requirements that is based on a particular dependency hierarchy for cognitive skills.

ConsScale defines the following levels that are considered relevant to the functional aspects of consciousness: 2 – Reactive, 3 – Adaptive, 4 – Attentional, 5 – Executive, 6 – Emotional, 7 – Self-conscious, 8 – Empathic, 9 – Social, 10 – Human-like, 11 – Super-conscious. For each level a number of abilities are listed. The ConsScale method does not address the phenomenal appearance of consciousness in any direct way.

It is obvious that the extent of consciousness, which is known to exist (like in humans), can be evaluated by ConsScale. However, the evaluation of artificial agents with ConsScale tells only something about the extent of cognitive capacities of the evaluated agent. The passing of the ConsScale tests does not prove that the artificial agent were phenomenally conscious to any degree in the first place.

20.5. The Ultimate Test for Existential Robot Consciousness

Let's assume that we have an advanced cognitive robot that is able to communicate using a natural language. We can talk about this and that and ask questions, but could there be any definite, absolute way to determine by conversation that the robot is conscious and self-aware? Yes, there is. The author has argued that there is one question that should show beyond all doubt that a robot is conscious [Haikonen 2003].

However, this question is not something that we should ask the robot. It would be a question that the robot asks us. This question should also be one that is neither prompted by us nor a product of any kind of programming. This question would also be at the root of everything.

This question would be: *Where did I come from?*

Chapter 21

An Experimental Robot with the HCA

21.1. Purpose and Design Principles

The Haikonen Cognitive Architecture (HCA) is a biologically inspired theoretical construct, which calls for empirical verification. However, there is challenge. HCA is designed to operate with meanings, and consequently it has to use self-explanatory sensory information for the grounding of meaning. This property cannot be simulated by computers as these do not accept self-explanatory information. Therefore a real hardware implementation experiment is necessary. For this purpose the author has designed a small Experimental Cognitive Robot, XCR-1 [Haikonen 2010, 2011].

The robot XCR-1 is a small (1.7 kg, ca. 3 lbs) autonomous, three-wheel robot with gripper arms and hands and simple visual, auditory, touch, shock and petting sensors, see Fig. 21.1. XCR-1 utilizes a simple version of the HCA, and is based on associative neural networks that are realized with hardware; they are not simulated by software.

XCR-1 has a set of motor action routines and innate reactions to stimuli. These routines and the innate reactions will combine in various ways depending on the situation. Cognitive control may override innate reactions and modify the robot's behavior.

XCR-1 is not a digital system. It does not utilize microprocessors, and is not program-driven, there is no program code what so ever. As such XCR-1 represents a completely different approach to autonomous self-controlled robots.

203

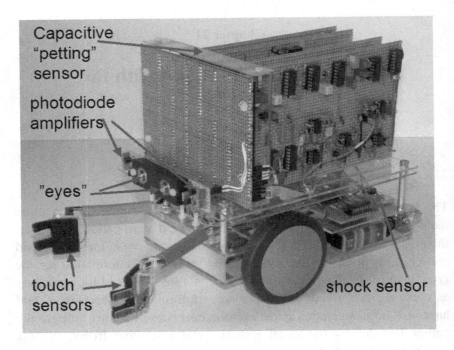

Fig. 21.1. The XCR-1 robot has three wheels for motion, gripper arms and hands with touch pressure sensors, and visual sensors, a microphone for sound perception, a shock sensor that detects mechanical shocks anywhere in the body and a petting sensor. The robot utilizes three small audio quality DC-motors.

21.2. Architecture

The robot XCR-1 utilizes a cognitive architecture that is a scaled-down version of the Haikonen Cognitive Architecture (HCA). The XCR-1 has the following main functions: Target search, detection, verbal identification, approach and gripping; Pain as dynamic disruptive system condition; Association of emotional values with objects; Motivation and reactions by emotional values; Match/mismatch detection and reporting; Simple natural language (English) verbal report (self-talk); Limited speech recognition; Limited verbal learning.

The block diagram of the robot XCR-1 is given in Fig. 21.2.

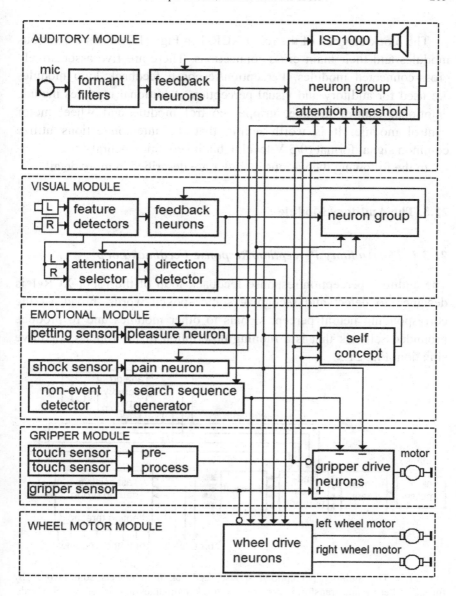

Fig. 21.2. The block diagram of the robot XCR-1 shows the auditory, visual, emotional, gripper and wheel motor modules and their cross-connections. Threshold control lines are not shown here.

The block diagram of the robot XCR-1 in Fig. 21.2 shows the various modules and their main cross-connections. There are five associatively cross-connected modules. Perception/response feedback loop modules are used for auditory and visual perception. Additional modules include emotional control module, gripper control module and wheel motor control module. It is worth noting that the interconnections utilize common signal format (0/5 V levels), but no common neural code.

In the following the various modules are described in more detail.

21.3. The Auditory Module

21.3.1. The Auditory Perception/Response Feedback Loop

The auditory perception/response feedback loop of the robot XCR-1 is designed to detect and recognize some spoken words, broadcast the corresponding neural percept signals to other modules and to produce grounded self-talk that is a running description of the robot's cognitive situation, Fig. 21.3.

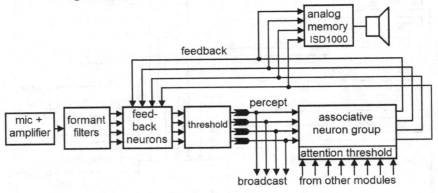

Fig. 21.3. The auditory perception/response feedback loop detects some words by formant filtering and transforms percepts from other modules into inner speech words, which are transformed into heard natural language words by the ISD1000 circuit.

A small electret microphone is used for the detection of sounds. The signal from the microphone is amplified and forwarded to formant filters that allow the detection of some spoken words. The detectable words generate patterns of neural signals, which are forwarded to the feedback neurons. The threshold circuit selects the actual percept signal pattern, which is broadcast to other modules.

The associative neuron group accepts broadcast percept signals from other modules via an attention threshold circuit. The accepted broadcast signals evoke associatively neural word signal patterns, which are fed back to the feedback neurons and are also forwarded to the speech producing analog memory circuit (ISD1000).

21.3.2. Speech Recognition

Spoken word detection is based on formant filters that detect the spectrum density of the sounds. Formant filters are narrow band-pass filters that have their band-pass ranges tuned to suit the formant frequencies of the human speech. These formant filters include a rectifying and smoothing circuit, which outputs a DC-voltage that indicates the signal intensity within the detected formant frequency band. Different combinations of formant frequencies correspond to different vowels. This process allows the recognition of the words regardless of their pitch.

In this simple application only vowel detection is used. This process is very limited, but suffices for the detection of few words and for the demonstration of the basic operational principles. It would be a straight forward task to extract additional auditory features by the addition of further formant filters, transient detectors and sequence detection, but this would increase the number of required neurons beyond easy discrete realization.

21.3.3. Speech Production

Speech production in XCR-1 is based on pre-recorded words that are stored in a non-volatile analog audio EEPROM chip, type ISD1000. (This type is now discontinued and replaced by the ISD1400 series devices.) This chip contains also a low-power audio amplifier that drives directly a small loudspeaker. The words are stored in this chip as temporal sequences of analog samples, and consequently no analog-digital and digital-analog conversions are used. Nevertheless, the memory locations are addressable with eight address lines. In principle, the address is a binary number and each address count equals 0.125 seconds of stored sound. In XCR-1 four most significant address bits are used; this gives a time period of 2 seconds. This is more than enough for one short word. The total capacity of the chip is 20 seconds, thus the maximum of ten words can be stored. The use of shorter time period would be possible and this would allow a larger vocabulary of short words.

It should be noted that the actual speech production method is of no theoretical significance here. Speech could also be produced by synthesis, but in the framework of this experiment it would only complicate the circuitry without providing any new insights. The important issue here is the demonstration of the grounding of meaning of the produced words.

21.4. The Visual Module

21.4.1. Visual Perception/Response Feedback Loops

The visual perception/response feedback loop module of the robot XCR-1 is designed for the purpose of searching and detecting active targets and their relative directions in respect to the robot heading.

The visual perception/response feedback loop module produces two kinds of information, namely the object feature information (identity) and the object direction information (location). The object feature information is broadcast to the auditory and emotional modules for naming and emotional significance evaluation, and the direction information is broadcast to the visual location memory and to the wheel motor module so that the robot can find its way to the object. The visual perception/response feedback loop module receives information from the auditory module and the emotional module. The visual perception/response feedback loop module is depicted in Fig. 21.4.

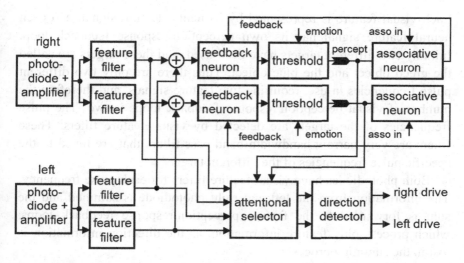

Fig. 21.4. The basic visual perception/response feedback loop module consists of two visual feature perception/response feedback loops, an attentional selector and direction detector. The direction detector provides information to the differential wheel drive circuits. Visual location memory is not shown here.

The visual perception/response feedback loop module utilizes two large area photodiodes (BPW34) as visual sensors. These are located in front of the robot and their distance is 30 mm from each other. Short focal length lenses (from cheap single use cameras) project the images of the targets on the photodiodes allowing improved detection and direction sensitivity. Each photodiode has its own amplifier.

The circuitry of visual perception/response feedback loop module consists of two visual feature perception/response feedback loops, an attentional selector and direction detector. The direction detector determines the direction of the attended object. The detailed operation of these circuits is described in the following.

21.4.2. Object Recognition

This robot is designed to detect visually active target objects. The active targets transmit pulsed infrared light and have different pulse frequencies. These frequencies constitute the detected visual features.

Each visual feature is represented by its neural feature signal, and each neural feature signal has its own perception/response feedback loop. Initially two different targets have been used, and these targets are called the green object and the blue object. Thus there are only two different pulse frequencies in use, requiring two feature signals, and therefore the number of visual perception/response feedback loops is two. The pulse frequencies of the targets are detected by visual feature filters. These filters are very narrow bandwidth band-pass filters that are tuned to the specific pulse frequencies of the different targets.

Both photodiodes have visual feature filters for each pulse frequency. The information from the left and right photodiode is summed and the sum is forwarded to the feature perception/response feedback loops, which process only feature information, as the direction information is lost in the summing process.

21.4.3. Direction Detection and Differential Drive

Visual direction detection is based on the parallax effect, and it is used for differential motor drive, see Fig. 21.5.

When the pursued object is directly in front of the robot, both photodiodes generate equal signals. If the object is on the left side, then, due to the parallax effect, the projected image of the target falls only partially on the right photodiode and the produced signal is weaker. At the same time, the projected image of the target covers more of the sensitive area of the left photodiode, and the produced signal is stronger. In principle, these signals could be amplified to drive directly the wheel motors; right signal would drive the left motor and vice versa. For instance, if the object is on the left side, then the left signal would be stronger and the right motor would run faster. This would turn the robot towards left, and eventually the object would be directly in front of the robot. At that moment both photodiode signals would be equally strong, and both wheel motors would run with the same speed. In this way a closed feedback control will arise whenever the robot is approaching a target. This feedback control loop ensures that the robot will eventually reach a position, where the target is quite exactly between the robot's gripper hands.

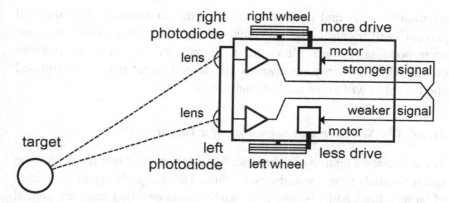

Fig. 21.5. The principle of driving towards a target with differential drive. Due to the parallax effect the projected image of the target falls only partially on the right photodiode and the produced signal is weaker. The stronger signal from the left photodiode drives the right wheel and the weaker signal from the right photodiode drives the left wheel. Consequently, the robot turns towards the target. In practice a certain improvement is required, see text.

However, this direct method has a drawback. When the detected object is far away, the photodiode signals are very weak, and the robot would move very slowly if at all. Therefore the distance information and the direction information must be separated from each other, and only the direction information that does not depend on the distance, must be used to control the wheel motors. This is done by the direction detector circuit that compares the relative strengths of the photodiode signals, and based on these, produces full amplitude wheel motor drive signals regardless of the distance of the target. Similar principle can be used for the sound direction detection.[2]

The robot may be seeing more than one target (in this case two) at the same time. Therefore an attentional selection mechanism is necessary. Artificial "saccades" allow the initial focusing of attention on each object in turn. The visual perception/response feedback loops are able to determine, which object is to be attended to, and this is where the final decision is made. Attention in the visual module is controlled by

---

[2] This is demonstrated in the "robot sound direction detection" demo video at https://www.youtube.com/user/PenHaiko

emotional values and possible external verbal commands. The attended percept will control the selector, which will then pass only the corresponding feature filter output signals to the direction detector circuit. Thus the direction towards the desired target will be determined and the robot will move towards that target.

21.4.4. The Selection and Searching of a Target

The associative neurons of the visual perception/response feedback loop receive signals from the auditory module. These signals depict the names of targets. Each name is able to evoke the corresponding feature signal at the feedback loops. This constitutes a virtual "imagined" visual percept of the corresponding target. This imaginary percept executes priming; if the same target is now actually detected, then the percept signal level will be elevated and this target will be favored. The selected percept feature controls also the direction detection and consequently the direction towards the named target will be determined.

A target may be selected also by its emotional value. The emotional system controls the threshold levels at the outputs of the feedback neurons. The threshold level for a "good" feature is lowered and consequently the "good" target is favored over a neutral one.

21.4.5. Match/Mismatch Detection

Match and mismatch conditions are detected within the visual and emotional perception loops. The internally evoked percepts (feedback patterns) are automatically compared with the actually sensed percepts (feature patterns), and the corresponding match and mismatch signals are generated. Global match and mismatch signals are derived from the individual match and mismatch signals from the visual and emotional perception loops. These global match and mismatch signals are associated with the words "yes" and "no" that can be uttered by the robot, when the match and mismatch conditions occur. In this way a verbal report of the match and mismatch conditions can be had.

21.5. The Emotional Module

21.5.1. Pain and Pleasure

The emotional system of the robot XCR-1 is based on the concepts of "pain" and "pleasure". "Pain" is a dynamic attention disrupting system condition that refocuses attention and leads to withdrawal and avoidance reactions as well as fast emotional learning. "Pleasure" is a condition that leads to the focusing of attention to the pleasure-producing situation, which will then be pursued and sustained. In XCR-1 pain and pleasure are not symbols or numeric values. Instead, they are observable, nameable and reportable sub-symbolic system reaction conditions.

In XCR-1 pain reactions are initiated by shock vibrations in the robot frame. These vibrations are detected by a shock sensor, which is bolted to the body of the robot. This sensor consists of a miniature magnetic earphone (from a toy), an amplifier and a threshold circuit. Due to band-pass filtering the shock sensor is not sensitive to normal external sounds or to the rather weak sounds of the motions of the robot itself.

The hard-wired pain reactions in the robot include the release of any gripped object, backing off or reversing the motion direction and dynamic disruption of attention (pain modulation). If the robot stands still and is hit from behind, the robot will jump forward a little bit.

A "petting sensor" on top of the robot is used to initiate pleasure conditions. This sensor is sensitive to touch.

The signals from the "petting" and shock sensors are forwarded to "pleasure" and "pain" neurons. These neurons are used to associate positive emotional value, "pleasure", and negative emotional value, "pain", with visual percepts. The robot can report verbally positive and negative emotional values by the words "good" and "bad". Active pain is associated with the word "hurt".

21.5.2. Emotional Significance

Petting and hitting can be used to change the instantaneous behavior of the robot by the evoked emotional reactions. Better overall behavior would follow, if the applied petting and hitting, reward and punishment, could be remembered and associated with the corresponding activity. This has been implemented in XCR-1. The robot can be taught and can learn by itself, which actions are desirable and worth pursuing, and which ones are to be avoided. The robot can accumulate a track of positive and negative emotional significance, of course within its limitations.

In XCR-1 emotional significance is facilitated by the pleasure and pain neurons. These neurons associate pain and pleasure with the objects that are related to pain and pleasure producing events. In this way these objects will be accompanied by emotional values, which will change the behavior of the robot by evoking the corresponding system reactions, when the object is encountered. The emotional values will also act as motivational factors before the objects are encountered; the robot will search objects with positive emotional value, and avoid objects with negative value.

21.5.3. Non-Events

The emotional module contains a "non-event" detector, which senses extended periods of non-activity. This corresponds loosely and functionally to the feeling of boredom when nothing is happening. When a non-event state is detected, the forward-turn-forward-turn search routine is initiated; the robot will begin to run around and seek targets. If the robot has been gripping an object during the non-event state, the gripper will release and the robot will back off before initiating the search routine.

21.5.4. Self-Concept

Sensory percepts can relate to the objects, their relationships and actions of the external world or to the perceiving subject itself, to its own body, own actions, own feelings and thoughts. The percepts of the subject itself allow the creation of a self-concept, which is a necessary requirement for self-consciousness. This requirement has been recognized, for instance, by Damasio and his proposition is that the basic sense of self is grounded to the representations (percepts) of the body [Damasio 2000, p. 22].

In XCR-1 the percepts of motion, touch and "pain" are used to evoke a self-concept. It is the robot itself that is moving, it is the robot itself that feels the touch, and it is the robot itself that is in pain; the presence of these percepts is always related to the experiencing self, which is the common denominator for these percepts. In XCR-1 the evoked denominator is in the form of a self-concept signal, which is associated with the aforementioned things of "self". This signal allows also the association and use of a verbal self-symbol, in this case the word "me". The robot may utter: "me hurt", when hit, "me search", when searching for an object, and "me touch", when the touch sensors indicate contact. These utterances may be followed by the name of the searched or touched object. Current wiring and vocabulary in XCR-1 do not allow the reports like "me see (something)" or "me hear (something)", but these would be simple additions along the already implemented lines.

21.6. The Wheel Drive Module

The robot XCR-1 has a two wheel differential drive system. The wheels are originally rubber rimmed furniture wheels with ball bearings. Each wheel has its own DC-motor, and the shaft of the motor drives directly the wheel's rubber rim. This is a simple and compact arrangement that requires no gearboxes resulting in very silent overall operation.

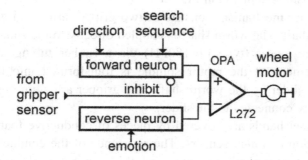

Fig. 21.6. The wheel motor drive circuit. The circuit is similar for both motors.

The wheel motor drive circuit for each wheel is given in Fig. 21.6. The wheel motors (type FF-180PH) are silent audio quality cassette load motors from surplus car C-cassette player mechanisms. These motors have good torque and only 50 mA no-load current, therefore they can be easily driven directly by one ampere-rated power operational amplifiers (such as L272).

The wheel motors are controlled by forward and reverse neurons. These neurons cause the power operational amplifier to drive positive or negative voltage to the motor, which will then run forwards or backwards. The forward neurons are activated by the direction signals from the visual direction detector and the sequential search routine signals that are a response to a detected non-event state. The gripper sensor outputs a signal, when a target is located right between the gripper hands. At that moment any forward motion must be stopped to avoid collision and to allow the gripper hands to grip the target properly. This is achieved by inhibiting the output of the forward neuron by the gripper sensor signal. However, the gripper sensor signal does not inhibit a possible reverse motion.

Reverse motion is caused by the activation of the reverse neurons. These neurons receive signals from the emotional module.

21.7. The Gripper Module

The gripper mechanism allows the robot to grip suitably sized objects. The mechanism is depicted in Fig. 21.7.

The gripper mechanism consists of two gripper arms that are driven by a worm shaft. The worm shaft is coupled directly to the spindle of the gripper drive motor (type FF-180PH) via a rubber tubing joint. The rotary movement of the motor spindle is transformed into the linear motion of a slider by the worm shaft. The gripper arms are operated by levers that are connected to the slider.

The gripper hands are covered by pieces of conductive foam plastic, which function as touch sensors. The resistance of the conductive foam depends on the applied pressure, and this can be easily sensed and translated into a corresponding voltage value by a high-impedance sense amplifier. In XCR-1 on/off touch sensing is used by applying a threshold circuit at the output of the sense amplifier. (Black conductive foam plastic sheets can sometimes be found in semiconductor packages. There are two varieties, brittle and soft. The latter is useful in this application.)

gripper mechanism

gripper motor

gripper motor driver (power OPA)

power conditioning

Fig. 21.7. The gripper mechanism. The rotary movement of the motor spindle is transformed into the linear motion of a slider by the worm shaft. The gripper arms are operated by levers that are connected to the slider.

An optical sensor (gripper sensor) detects objects between the gripper hands. When an object is detected between the gripper hands, the robot will stop, if it has been moving. Next, the gripper hands will close, unless this reaction is overridden by cognitive control. The gripper motor will stop, when the touch pressure at both hands reaches a set limit. At that point the gripper holds the object quite firmly, but not too hard. There are limit switches at both ends of the worm shaft drive (these are micro switches from a dead PC mouse). These switches stop the gripper motor when the worm shaft drive reaches its extreme positions. The principle of the gripper circuit is shown in Fig. 21.8.

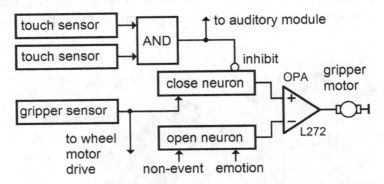

Fig. 21.8. The gripper circuit. The power operational amplifier (OPA) drives positive or negative voltage to the motor. Positive voltage closes the gripper and negative voltage opens it.

The gripper circuit consists of two neurons that are connected to a power operational amplifier (OPA). The excitation of the "close neuron" causes the OPA to drive positive voltage to the gripper motor and this will close the gripper arms. Touch sensor signals will inhibit the "close neuron" and the gripper motor will stop. The "open neuron" is an OR neuron and is controlled by the non-event detector and negative emotion. The excitation of the "open neuron" causes the OPA to drive negative voltage to the gripper motor and this will open the gripper arms.

21.8. Self-Talk

The self-talk in the XCR-1 robot is a running report of the instantaneous content of the neural system. The self-talk consists of a limited number of natural language (English) words that are produced by the ISD1000 analog audio memory chip in the previously described way by translating the signal patterns in the auditory module into the binary addresses for the ISD1000 chip. The auditory signal patterns inside the auditory module are evoked associatively by broadcasts from the other modules, and therefore act as symbols for the activity in the other modules. The

SD1000 chip translates the neural signal pattern symbols into natural language words with meanings that are grounded to the active percepts of the various modules. Therefore the self-talk offers a window into the inner workings of the cognitive system of the XCR-1.

The various modules of the XCR-1 robot broadcast continuously and simultaneously their percept signals to the other modules and also to the auditory module. However, only one word can be uttered at a time and therefore the broadcasts must be accepted serially, only one broadcast at a time. This function is realized by an attention threshold circuit at the associative input of the auditory module.

Several possibilities for this kind of attentional selection are available. The word order could be determined by inner models and emotional values or a fixed word order may be used. Flexible word orders would be desirable in more complicated systems. In this simple robot fixed word order works quite satisfactorily, and is easy to implement with minimal hardware. Here the attention threshold circuit is made to "scan" the incoming broadcasts sequentially and accept only one broadcast at a time. This broadcast signal will then evoke the corresponding word. The implemented word order corresponds to the order of subject – verb – object – adjective.

New spoken words cannot be initiated before the completion of the current word, therefore the attention threshold circuit can accept the next broadcast only after the completion of the previous word. This timing is facilitated by the end-of-word signal provided by the ISD1000 chip. Fall-back timing is used when no words are uttered and consequently no end-of-word signals are available.

The self-talk begins as soon as the XCR-1 perceives something. The percepts of the "green" and "blue" test targets elicit word sequences like "me search green" or "me search blue", when the robot starts to move towards the detected target. ("Green" and "blue" are the names of the test targets. For the benefit of an observer, the targets have green and blue bands, and in the robot the visual perception of "green" and "blue" targets are indicated by green and blue leds; the robot itself does not utilize these features.) When a target has been captured between the gripper hands, a word sequence like "me touch green" is produced. Hitting the robot will elicit the comment "me hurt". Emotional values can be associated with the test targets. These will we reported as "green bad" or "green good".

The simple self-talk of the XCR-1 robot reports the activity of all modalities in a symbolic way within the auditory modality. In doing so, the self-talk of XCR-1 illuminates the practical issues of basic grounding of word meaning. Complete natural language is more than that, but without this first step there cannot be any language. The next experimental steps would be the implementation of advanced sentences and the utilization of self-talk in planning and reasoning according to the multimodal model of language [see Haikonen 2003, 2007].

21.9. Experiments and Tests with XCR-1

The robot XCR-1 is able to display different behaviors in different situations. These behaviors are based on a few hard-wired reactions, learned associations and on the associative information integration of the HCA architecture. These behaviors do not result from an execution of any program code, as there is none in the XCR-1. Each behavior can be demonstrated by subjecting the robot XCR-1 to the desired situation. No internal settings for different responses are done, and besides, there are no provisions for such adjustments. These behaviors have been demonstrated by various experiments and short demo videos of these are available[3]. Here some of these experiments are described.

21.9.1. Visual Stimuli Response

XCR-1 can recognize two different target objects, "blue" and "green". When the robot perceives e.g. the green target, it turns towards the target, and the name and the good/bad/neutral emotional value of the target is evoked. If the target is deemed to be neutral or good, the robot will home in on it and report "me search green". Eventually the robot will grab the target, and the report will be "me touch green". If the target is deemed to be bad, then instead of approaching, the robot will back off and will

---

[3] Demo videos of the robot are here: https://www.youtube.com/user/PenHaiko

report "green bad". The main activated connections during a response to a seen object are depicted in Fig. 21.9.

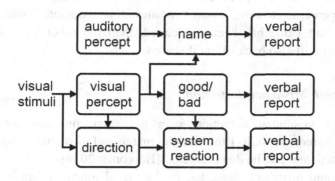

Fig. 21.9. Main activated connections during a response to a seen object.

21.9.2. *Verbal Evocation of a Virtual Visual Percept*

The XCR-1 is able to understand some spoken words, like the names of the targets ("green" and "blue") and the words "bad" and "hurt". Heard target names evoke virtual visual percepts of the corresponding targets. The presence of the evoked virtual visual percept can be verified by an oscilloscope and by its behavioral consequences. For instance, the virtual percepts of targets evoke the emotional good/bad values of the targets, and this is verbally reported. The evoked emotional values also evoke the corresponding system reactions, like evasion. This test demonstrates also that the robot understands the meanings of the heard words.

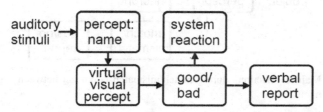

Fig. 21.10. Main connections during a response to a heard name. The name evokes a virtual percept with its emotional value and related system reactions.

Figure 21.10 depicts the main connections during a response to a heard name. During this test the name of one of the target object is spoken. The name-percept evokes the virtual visual percept of the object, and this percept, in turn, evokes whatever is associated with it. If emotional bad value has been associated, then the robot will verbally report it and will also back off as if scared.

21.9.3. Match/Mismatch Detection

Match and mismatch detection is a necessary function for every cognitive system. In XCR-1 multimodal match and mismatch detection is implemented, and can be demonstrated [Haikonen 2014].

Match and mismatch detection in the visual modality can be tested and demonstrated by showing and verbally naming one of the green and blue target objects simultaneously, see Fig. 21.11. The heard name will evoke the corresponding virtual visual percept in the visual perception/response loop. If the evoked virtual percept is similar to the actually seen object, match condition occurs, and a verbal "yes" report results. If the evoked virtual percept does not match the actually seen object, a "no" report is generated. This test also demonstrates that the robot understands the meanings of the heard names.

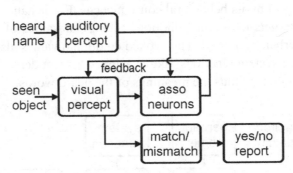

Fig. 21.11. Main connections during match/mismatch condition between a seen object and a heard name.

Match and mismatch detection in the emotional modality can be tested and demonstrated in a similar way, see Fig. 21.12. The robot can be asked "(does it) hurt". Normally the robot will answer "no". If the robot has been previously hit and the pain condition still prevails, the robot will answer "yes me hurt". In this response the word "yes" is generated by the match/mismatch condition, "me" is generated by the self-concept circuit, and the word "hurt" is generated by the pain circuit. This test demonstrates also that the robot understands the meaning of the word "hurt".

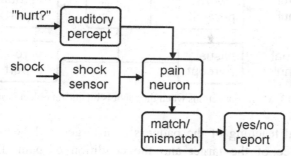

Fig. 21.12. Match/mismatch response generation during the question "does it hurt".

21.9.4. Pain and Emotional Bad Value

Pain effects and emotional value learning can be tested in various ways. When XCR-1 is hit, the mechanical shock is detected and transformed into corresponding electric signal. This signal leads to global disruptive attention threshold modulation ("actual pain") and to the generation of the corresponding verbal "me hurt" report.

A target can be associated with pain. This can be done in the following ways: 1) hitting the robot while the target is present and seen, 2) hitting the robot, while the virtual visual percept of the target is present, 3) by repeated verbal teaching. In each case the actual and virtual visual percepts of the target will lead to evading action and a verbal report like "green bad".

The actual hitting of the robot leads to the instantaneous association of "bad" with the seen object ("emotional learning"). After this event the robot will try to avoid the object whenever it sees it. The robot will also report that the object is bad, see Fig. 21.13.

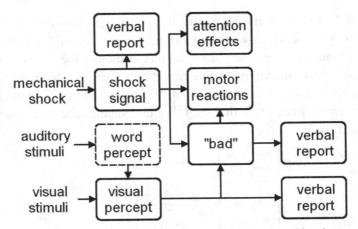

Fig. 21.13. Main connections when the robot is hit and an object is seen.

Full verbal teaching of the badness of a target is done without the actual presence of the target and the condition of pain. During the teaching a sentence like "green is bad" is spoken repeatedly several times. (The verb "is" is ignored by the robot, but may be used for the benefit of an observer.) This situation calls for the understanding of the meaning of the consequent words, "green" and "bad". The word "green" evokes the corresponding virtual visual percept in the visual module, while the word "bad" evokes nothing there. Likewise, the word "green" evokes nothing in the emotional module, while the word "bad" evokes the emotional value "bad".

The mere instantaneous evocation of these virtual percepts for the duration of each word is not enough, as entities that are not present simultaneously cannot be associated with each other. Therefore these percepts must be sustained simultaneously for a while, so that the association can take place. This short-term memory action is an inherent property of HCA perception/response feedback loops, and is thus effected without any additional circuitry.

At first, the teaching of "green is bad" leads to mismatch, because this contradicts the previous knowledge in the robot; technically, "green" has not been associated with "bad". Consequently, the robot will report "no". With repetition, the pre-synaptic weights of the corresponding associative neurons will eventually exceed the threshold and the associative connection is made. The robot will now report "yes". The learning can be verified by presenting the green object and noticing that the robot will now report "green bad" and it will also try to avoid that object. The connections during verbal teaching are shown in Fig. 21.14.

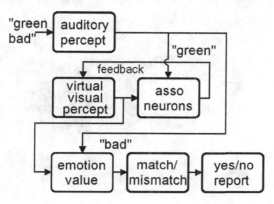

Fig. 21.14. Main connections during the verbal teaching that green is bad.

21.9.5. Short-Term Visual Location Memory

Visual location memory allows the association of seen objects with their seen locations, so that what is where can be known, also outside the focus of gaze direction. Without this memory situational awareness would be limited. Also, this kind of a memory is needed for imagination, for the assembly of complex imagery.

Visual location memory is a short-term memory, where the contents must be updated as necessary. There can be only one object at each memory location, and when the object is replaced with another, the memory must be updated.

A visual location is defined by the gaze direction. XCR-1 does not have a head that could be turned towards different directions in respect to the robot's body; the gaze direction is fixed and is always in line with the body. Therefore the robot must turn itself in order to change gaze direction. Unfortunately this also means that the robot's body cannot be used as a reference direction, instead an external reference must be used. In the experiments with XCR-1 this reference is an additional illuminated object that acts as an external reference point. The set-up for the testing of visual memory in XCR-1 is given in Fig. 21.15.

Fig. 21.15. The set-up for the testing of visual memory function. The external reference point is provided by the illuminated object between the "green" and "blue" target objects. The objects can be covered with a folded piece of cardboard.

In the visual location memory experiment the "green" and the "blue" target objects are placed on the left and right sides of the reference object, far enough that the robot does not see them without turning.

The visual memory function test begins with the showing of the target object locations. The robot is turned towards left, whereupon the robot immediately reports the seen object, "green". At this moment the location direction associates itself with the seen object. Then, the robot is turned towards right, and the same occurs, now with the other target object and direction.

The existence and operation of the location memories can be verified and tested by covering the target objects and turning the robot towards left and right. Each direction will evoke the virtual percept of the previously seen target object, and the robot will report the object's name. However, because the object cannot be seen, a mismatch condition occurs between the seen and expected percept, and the robot reports "no". The object is not where it was supposed to be. If the object is uncovered, the seen and expected percepts match, and the robot reports "yes". This verifies the proper operation of the short-term visual location memory and also again the correct operation of the visual module match/mismatch detection. The operation of the short-term visual location memory in XCR-1 is depicted in Fig. 21.16.

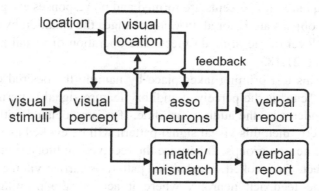

Fig. 21.16. The operation of the short-term visual location memory in XCR-1.

21.9.6. Visual Priming

Sensory priming gives an internal preference for sensory percepts. Visual priming is useful when certain objects are searched for, or when a certain object must be selected amongst many. Visual priming in XCR-1 can be demonstrated by using two target objects [Haikonen 2018]. The target objects (green and blue) are placed in front of the robot, see Fig. 21.17.

Fig. 21.17. The set-up for the visual priming test.

The "green" and "blue" target objects are placed at such distances from the center position that the robot will not react to them. However, the visual sensors will receive signals from these objects in a way of peripheral vision, but the signal level is not exceeding given thresholds, and consequently no percepts are formed and no responses are generated. The target objects are ignored. Priming changes the situation by elevating the signal level of the primed object. The operation of visual priming is given in Fig. 21.18.

In the this test priming takes place by naming the desired object. In Fig. 21.18 "cue" is the perceived auditory feature signal pattern (a heard name) broadcast by the auditory module. If it is associated with a visual signal pattern, then this visual signal pattern will be evoked as the output of the asso neuron groups. Otherwise the received auditory signal pattern has no effect. The evoked visual signal pattern is carried via the feedback loop to the feedback neurons, where it acts as the priming pattern. Feedback neurons receive also the sensory signals from the visual sensor, and each feedback neuron produces the linear sum of the intensities of the sensory signal and the priming signal.

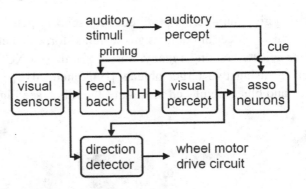

Fig. 21.18. The operation of visual priming in XCR-1.

Without priming the intensity of the sensory signals generated by peripheral vision would not exceed the set threshold value of the threshold circuit TH, and no percept would be produced. However, when the peripherally seen object matches the expected object, the sensory signals and priming signals match and sum in phase. This raises the feedback neuron output signal level above the threshold. The object will now be detected, it will be named and reported, and its location determined. As a result, the robot will turn towards and approach the named object.

In this visual priming test an external stimulus, a spoken name, is used as the priming cue. An internal stimulus, such as emotional good value could be used as well. In this way the robot could autonomously select between different objects.

21.10. Machine Consciousness and XCR-1

It is argued earlier in this book that consciousness is a phenomenon, where sensory neural activity appears as reportable qualities of the world instead of appearing as the actual neural activity or not appearing at all. Every moment the contents of consciousness consists of percepts and only percepts, real or virtual. These percepts have the form of qualia, self-explanatory qualities, possibly with associated meanings. When all percepts vanish, also consciousness vanishes, because there will be nothing to be conscious of. As such, consciousness is not an entity or agent. It is neither any material nor immaterial substance, it is just an inner appearance. But does XCR-1 have it?

It is presented here that the basic requirements for consciousness are 1) Sensory perception process, which produces self-explanatory sensory percepts in the form of qualia, 2) Feedback loops that allow the introspection of the mental content in terms of virtual percepts, 3) Cross-connections and short-term memory, for the generation of report.

Actual cognition calls for additional functions and associative symbol processing machinery for thinking and reasoning, but these are not a necessary prerequisite for basic phenomenal consciousness. The situation is the opposite; consciousness with self-explanatory percepts is necessary for the grounding of meaning of symbols.

The robot XCR-1 is based on the HCA architecture. The HCA architecture consists of sensory perception/response feedback loops that are able to import sensory information in self-explanatory forms. The perception/response feedback loops are able to introspect inner content in the form of virtual sensory percepts. The perception/response feedback loops provide also short-term memory function, which allows the reporting of the percepts to other perception/response feedback loops and also to outside, also in the form of verbal reports.

In this way the basic requirements for consciousness seem to be fulfilled in XCR-1. The decisive question is: Are the self-explanatory sensory percepts of XCR-1 comparable to natural qualia? In humans qualia are the way in which perception-related neural activity manifests itself as qualities of the world, instead of appearing as the actual neural activity. Does this happen in XCR-1? Does the "rigid stick" apply here?

In XCR-1 sensory information is technically conveyed by electric signal patterns. This information is self-explanatory; the patterns stand for the sensed qualities and properties, and are treated as the true or virtual presence of these. These patterns are not inspected as electric signal patterns, which remain only as the unobserved information carrier media. In this way these signal patterns operate like natural qualia, even though their internal appearance (if any) may not necessarily be the same. Qualia are self-explanatory information, but does all self-explanatory information have to appear internally as qualities of sensed entities? Perhaps, how else could it be self-explanatory? If so, then XCR-1 is conscious, but of course the scope and contents of this consciousness would be very, very limited. If the answer is no, then XCR-1 is a zombie, perceiving the world and operating with self-explanatory information like a true sentient being, yet without any internal appearances.

It should be obvious that the operational principle of HCA and the robot XCR-1 is radically different from digital computers that do not import external meanings in self-explanatory forms, and therefore fall victims to the symbol grounding problem. Also, digital computers operate with numbers. "Pain" in a digital robot is just a numeric value of some variable, and that does not hurt. In XCR-1 "pain" is not a numeric value or a symbol, it is a reportable dynamic disruptive system condition.

Digital computers are not conscious, and contemporary AI is not true intelligence. XCR-1 represents a radical attempt towards information processing with meanings, towards machines that understand their situation in their environment, machines that know what they are doing, machines that are conscious.

APPENDIX. Some Experimental Circuits

1. Simple Correlative Synapses

A correlative associative synapse learns the connection between two signals, s and a, via their repeated coincidences. Initially the synaptic weight w is zero and the synapse will not produce output regardless of the inputs. After learning the synaptic weight w goes to one, and the synapse will produce output, whenever the a-signal goes to one.

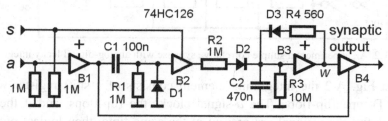

Fig. A.1. An artificial correlative associative synapse with analog correlator.

Figure A.1 shows the circuit diagram of an artificial correlative associative synapse with analog integrator for hands-on experimental purposes. This synapse is built around the CMOS quad 3-state buffer 74HC126. The R1, C1, D1 combination after the first buffer B1 generates more or less constant width narrow pulses as the response to

231

the *a*-signal. The buffer B2 is keyed by the *s*-signal, so that an output pulse is generated whenever the *s*-signal and the *a*-signal coincide. These pulses are accumulated by the leaky integrator, which is formed by the R2 and R3 resistors and the C2 capacitor. If the integrator receives a number of input pulses within a short period of time, the voltage of the C2 capacitor will exceed the threshold voltage of the buffer B4, and the synaptic weight *w* goes up and stays there due to the positive feedback via the current limiting R4 resistor and the diode D3. The synapse has now learned the associative connection between the *a*-signal and the *s*-signal, and thereafter the *a*-signal will evoke synaptic output. This synapse can be used in the demonstration of Pavlovian conditioning and also in simple robotic applications.

The leaky integrator of the circuit of Fig. A.1 has a limitation; the learning has to take place within a short period of time due to the short time constant of the R3C2 circuit. This shortcoming can be remedied by the use of digital or analog-digital integrators. An example of a simple correlative synapse with analog-digital leaky integrator is in Fig. A.2.

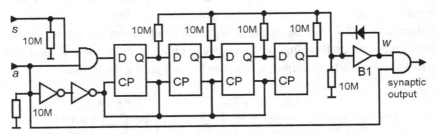

Fig. A.2. An experimental simple correlative synapse with analog-digital leaky integrator.

In Fig. A.2 the integrating element consists of a shift register with four D-type flip-flops. The *a*-signal clocks the flip-flops, and if the *a*-signal and the *s*-signal are present at the same time, then logical one is fed to the D-input of the first flip-flop, otherwise the D-input remains at logical zero. The same happens again at the next time, when the *a*-signal goes to one. Eventually a pattern of logical ones and zeros are shifted through the shift register. The outputs of the individual flip-flops are analogically summed together, and whenever three ones are present, the sum will exceed the input threshold of the buffer B1. This will then lock, and the synaptic weight w will be one. The synapse has now learned the associative connection between the a-signal and the s-signal, and thereafter the a-signal will evoke synaptic output. The integrator will not empty over time on its own, only repeated a-signals without the presence of s-signals can empty it. Therefore the learning can take place over on extended period of time.

2. Winner-Takes-All Threshold

The Winner-Takes-All Threshold (WTA) is used, when from a number of signals the strongest one is to be selected. There are several ways to do it. Here one example is given, where only one reference line between the signal channels is required. This circuit is given in Fig. A.3.

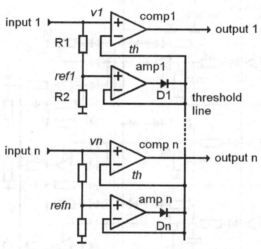

Fig. A.3. Winner-Takes -All circuit.

In Fig. A.3 the comparator comp1 compares the input voltage $v1$ to the threshold line voltage th. The voltage division network of R1 and R2 produces a reference voltage $ref1$, which is slightly lower than the input voltage $v1$. If the threshold line voltage th is lower than $ref1$, then the amp1 operates as a voltage follower, and outputs the $ref1$ voltage to the threshold line. This voltage will be slightly lower than the input voltage $v1$, and therefore the comp1 will produce logical one as its output.

If the threshold line voltage is higher than the *refl* voltage, then the diode D1 will be reversebiased, and the output of amp1 will go to zero having no effect on the threshold line voltage. The comp1 will now output zero, as the voltage at the inverting input is higher than the voltage *v1* at the non-interting input. It can be seen that the strongest input signal will be chosen, when a number of these circuits are connected together via the threshold line, winner takes all.

3. An Associative Neuron with Interference-Free Synapses

The previous circuits can be combined into neurons with several synapses, see Fig. A.4.

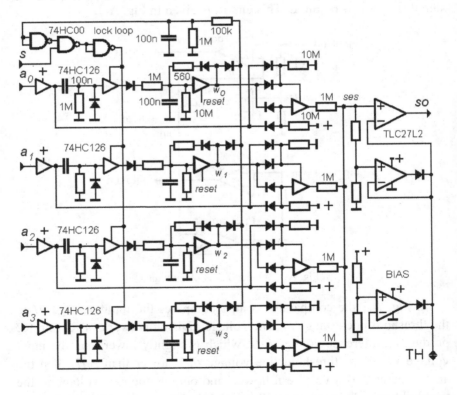

Fig. A.4. An experimental associative neuron with interference-free synapses. The lock loop prevents further learning as soon as one or more synaptic weights *w* go up.

The neuron circuit of Fig. A.4 will learn the association between the s signal and a pattern $\{a_0, a_1, a_2, a_3\}$ via repeated coincidences. In this case the meaning of the output signal so is the same as that of the s signal.

If s is set permanently high, then the neuron will learn a repeating pattern $\{a_0, a_1, a_2, a_3\}$ on its own. In that case the output signal so indicates only the presence of the learned pattern.

In this neuron circuit the interference-free synaptic rule is implemented with diode logic combined with the 3-state buffer 74HC126, see Fig. A.5.

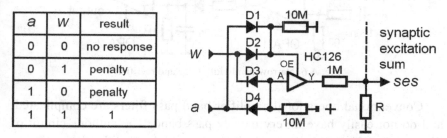

a	w	result
0	0	no response
0	1	penalty
1	0	penalty
1	1	1

Fig. A.5. The realization of the interference-free synaptic rule.

The first condition of the rule, no response if a and $w = 0$, is realized by the high impedance output state of the 3-state buffer 74HC126. The buffer is enabled only when OE is high, and this will not happen, if both a and w are low. In that case the output Y of the 74HC126 is in the high impedance state. Consequently the 1M resistor will not be connected anywhere, and therefore does not contribute to the synaptic excitation summing resistor network.

It can be seen that due to the diodes D1 and D2 the OE signal will go high, if a or w or both are high. In those cases the output Y of the 74HC126 has the same value as its input A. The value A is determined by w and a via the diodes D3 and D4, and can be high only when both a and w are high. In this case a positive value is added to the synaptic excitation sum ses. Otherwise the 1M resistor is coupled to ground and ses will be lowered. This corresponds to the penalty of the synaptic rule.

It should be obvious that the function of the circuit in Fig. A.5 can be realized in several different ways.

4. Narrow Bandwidth Filter

Even in simple auditory perception some kind of audio spectrum analysis is necessary. This can be done by using a group of very narrow bandwidth filters. An example of a very narrow bandwidth filter using an operational amplifier is given in Fig. A.6.

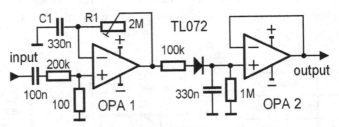

Fig. A.6. Narrow bandwidth filter and rectifier.

Conventional operational amplifier band pass filters are complicated, and do not easily have a very narrow pass band. The author's circuit of Fig. A.6 is different; it is simple, it has very high amplification and very narrow bandwidth. The pass band center frequency is determined by C1, R1 and the upper frequency limit of the operational amplifier OPA1. A low noise temperature compensated operational amplifier is required, for example the TL072, which works well when dual voltage supply is used. R1 should be a good quality multi-turn trimmer potentiometer for easy and exact tuning.

In Fig. A.6 the sine wave output from OPA1 is rectified and smoothed, and the output of OPA2 is a DC-voltage that is proportional to the amplitude of the sine wave.

5. Sound Direction Detector

Sound direction detection in robotic auditory perception using two microphones can be based on the phase difference or the intensity difference between the microphone signals. Phase difference is caused by the different distances between the two microphones and the sound source. Intensity difference is caused by the shadowing effect of the robot head. Sound direction detection using phase difference is effective at lower frequencies, while intensity difference is useful at higher frequencies. Figure A.7 gives an example of a simple intensity difference sound direction detector using operational amplifiers.

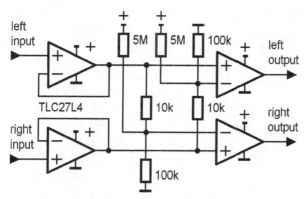

Fig. A.7. Binary output direction finder.

The sound direction detector of Fig. A.7 separates the direction information from the intensity information; the distance to the sound source does not affect the determination of the direction. The intensities of the left and right input signals are compared against each other. If the left input signal is stronger than the right input signal, the left output goes up and the right output remains at zero, and vice versa. The 5M resistors provide a small bias for preventing output, when inputs are zero. Operational amplifier types like TLC27L4, which can sense down to ground, must be used here when single supply voltage is used. A filter-rectifier combination, like that of Fig. A.6, should be used in front of this circuit.

It is possible to have additional comparators between the ones in Fig. A.7 for higher resolution direction detection.

Bibliography

Abbott, L.F. [1999] Lapique's introduction of the integrate-and-fire model neuron (1907), *Brain Research Bulletin* **50** (5/6), 303–304.

Aleksander, I. [2009] Essential Phenomenology for Conscious Machines: A Note on Franklin, Baars and Ramamurthy: "A Phenomenally Conscious Robot", *APA Newsletter on Philosophy and Computers* **8**(2).

Aleksander, I. and Dunmall, B. [2003] Axioms and Tests for the Presence of Minimal Consciousness in Agents, in O. Holland (ed.), *Machine Consciousness* (Imprint Academic), pp. 7–18.

Anderson, J. A. [1995] *An Introduction to Neural Networks* (MIT Press).

Arrabales, R. Ledezma, A. and Sanchis, A. [2009] "CERA-CRANIUM: A Test Bed for Machine Consciousness Research", in *Proc. International Workshop on Machine Consciousness 2009* (Hong Kong).

Arrabales R., Ledezma A. and Sanchis A. [2010] The Cognitive Development of Machine Consciousness Implementations, *IJMC* **2**(2), 213–225.

Baars, B. J. [1988] *A Cognitive Theory of Consciousness* (Cambridge University Press).

Baars, B. J. [1997] *In the Theater of Consciousness* (Oxford University Press).

Baars, B. J. and Franklin, S. [2009] Consciousness is computational: The LIDA model of Global Workspace Theory, *IJMC* **1**(1), 23–32.

Balduzzi, D. and Tononi, G. [2009] Qualia: The Geometry of Integrated Information, *PLoS Comput Biol* **5**(8): e1000462. doi:10.1371/journal.pcbi.1000462.

Bergson, H. [1988] *Matter and Memory* (Zone Books).

Block, N. [1995] On a Confusion about a Function of Consciousness, *The Behavioral and Brain Sciences* **18**(2), 227–287.

Boltuc, P. [2009] The Philosophical Issue in Machine Consciousness, *IJMC* **1**(1), 155–176.

Botvinick, M. and Cohen, J. [1998] Rubber hands `feel' touch that eyes see, *Nature* **391**, 756.

Brooks, M. [2014] *At the Edge of Uncertainty* (Profile Books UK).

Browne, W. N. and Hussey, R. J. [2008] Emotional Cognitive Steps Towards Consciousness, *IJMC* **1**(2), 203–211.

Chalmers, D. J. [1995a] The Puzzle of Conscious Experience, *Scientific American*, **237**(6), 62–68.

Chalmers, D. J. [1995b] Facing Up to the Problem of Consciousness, *JCS* **2**(3), 200–219.

Chella, A. [2008] Perception Loop and Machine Consciousness, *APA Newsletter on Philosophy and Computers* **8**(1), 7–9.

Corkill, D. D. [1991] Blackboard Systems, *AI Expert* **6**(9), 40–47.

Crick, F. [1994] *The Astonishing Hypothesis* (Simon & Schuster), p. 3.

Crick, F. and Koch C. [1992] The problem of Consciousness, *Scientific American* **267**(3), 152–160.

Damasio, A. [2000] *The Feeling of What Happens* (Vintage).

Dennett, D. [1987] Cognitive wheels: The frame problem of AI, in C. Hookway (ed.), *Minds, machines and evolution* (Baen Books), pp. 41–42.

Dennett, D. [2001] Are we explaining consciousness yet? *Cognition* **79**, 221–237.

Duch, W, Oentaryo R. J. and Pasquier M. [2008] Cognitive architectures: where do we go from here? in P. Wang, B. Goertzel and S. Franklin (eds.), *Frontiers in Artificial Intelligence and Applications, Vol. 171* IOS Press, pp. 122–136.

Fodor, J. [1975] *The Language of Thought* (Crowell).

Gallup, G. Jr., [1970] Chimpanzees: Self-recognition, *Science* 167, 86–87.

Gamez, D. [2008] *The development and analysis of conscious machines*, Ph.D. Thesis. University of Essex, Computing Department. p. 25.

Gibson, J.J. [1966] *The Senses Considered as Perceptual Systems* (Houghton Mifflin).

Gregory, R. L. [1998] *Eye and Brain* (Oxford University Press), p.168.

Haikonen, P. O. [1999] *An Artificial Cognitive Neural System Based on a Novel Neuron Structure and a Reentrant Modular Architecture with implications to Machine Consciousness*, Doctoral Thesis. Series B: Research Reports B4. Helsinki University of Technology, Applied Electronics Laboratory; 1999.

Haikonen, P. O. [1999b] Finnish patent no 103304.

Haikonen, P. O. [2003] *The Cognitive Approach to Conscious Machines* (Imprint Academic).

Haikonen, P. O. [2005] "You Only Live Twice: Imagination in Conscious Machines", in *Proc. Symposium on Next Generation approaches to Machine Consciousness: Imagination, Development, Inter-subjectivity, and Embodiment (AISB05)* pp. 19–25.

Haikonen, P. O. [2007] *Robot Brains; circuits and systems for conscious machines* (John Wiley & Sons).

Haikonen, P. O. [2009] Qualia and Conscious Machines, *IJMC* **1**(2), 225–234.

Haikonen, P. O. [2010] "An Experimental Cognitive Robot", in A. V. Samsonovich, K. R. Johannsdottir, A. Chella and B. Goertzel (eds.), *Biologically Inspired Cognitive Architectures 2010* (IOS Press) pp. 52–57.

Haikonen, P. O. [2011] XCR-1: An Experimental Cognitive Robot Based on an Associative Neural Architecture, *Cognitive Computation* **3**(2), 360–366.

Haikonen, P. O. [2014]. Yes and No: Match/Mismatch Function in Cognitive Robots. Cognitive Computation. Volume 6, Issue 2 (2014), pp. 158–163. doi: 10.1007/s12559-013-9234-z.

Haikonen, P. O. [2018] "Visual Priming in a Biologically Inspired Cognitive Architecture", in *Biologically Inspired Cognitive Architectures 2018 Proceedings of the Ninth Annual Meeting of the BICA Society* pp. 113–118.

Harnad, S. [1990] The Symbol Grounding Problem. Physica D 42: 335–346.

Harnad, S. [1992] The Turing Test Is Not a Trick: Turing Indistinguishability Is a Scientific Criterion, *SIGART Bulletin* 3(4), 9–10.

Harnad, S. and Scherzer, P. [2007] "First, Scale Up to the Robotic Turing Test, Then Worry About Feeling", in *AI and Consciousness: Theoretical Foundations and Current Approaches*, AAAI Fall Symposium Technical Report FS-07-01. pp. 72–77.

Hayes-Roth, B. [1985] A Blackboard Architecture for Control, *Artificial Intelligence* 26(3), 251–321.

Hebb, D. O. [1949] *The Organization of Behavior* (Wiley).

Hesslow, G. [2002] Conscious thought as simulation of behaviour and perception, *Trends in Cognitive Sciences* 6(6), 242-247.

Hinton, G. E., McClelland, J. L. and Rumelhart, D. E. [1986] Distributed representations, In D. E. Rumelhart, and J. L. McClelland (eds.), *Parallel Distributed Processing: Explorations in the Microstructure of Cognition, Vol. 1: Foundations* (MIT Press), pp. 77–109.

Holland, O. and Goodman, R. [2003] Robots with Internal Models: A Route to Machine Consciousness? in O. Holland (ed.), *Machine Consciousness* (Imprint Academic), pp. 77–109.

Holland, O., Knight, R. and Newcombe, R. [2007] A Robot-Based Approach to Machine Consciousness, in A. Chella and R. Manzotti (eds.), *Artificial Consciousness*. (Imprint Academic), pp. 156–173.

Hume, D. A. [2000] *Treatise of Human Nature*, ed. D. F. Norton and M. J. Norton (Oxford University Press).

Jain, A. K., Mao, J. and Mohiuddin, K. M. [1996] Artificial Neural Networks: A Tutorial, *Computer* 29(3), 31–44.

Jackson, F. [1982] Epiphenomenal Qualia, *Philosophical Quarterly*, 32, 127–36.

Kanerva, P. [1988] *Sparse Distributed Memory* (MIT Press).

Kawamura, K. and Gordon, S. [2006] "From intelligent control to cognitive control", in *Proc. 11th International Symposium on Robotics and Applications* (ISORA), (Budapest, Hungary).

Kinouchi, Y. [2009] A Logical Model of Consciousness on an Autonomously Adaptive System, *IJMC* 1(2), 235–242.

Koch, Ch. and Tononi, G. [2011a] A Test for Consciousness, *Scientific American* 304(6), 26–29.

Koch, Ch. and Tononi, G. [2011b] Testing for Consciousness in Machines, *Scientific American Mind* 22(4), 16–17.

LeDoux, J. [1996] *The Emotional Brain* (Simon & Schuster).

Lesser, E., Schaeps, T., Haikonen, P. and Jorgensen, C. [2008]. "Associative neural networks for machine consciousness: Improving existing AI technologies", in *Proc. of IEEEI 2008*, pp. 011–015.

Levine, J. [1983] Materialism and qualia: the explanatory gap, *Pacific Philosophical Quarterly* 64: 354–361.

Lewis, M., Sullivan, M. W., Stanger, C. and Weiss, M. [1989] Self-development and self-conscious emotions, *Child Development* **60** (1), 146–156.

Lieto, A. [2017] "The Explanatory Problems of Deep Learning in Artificial Intelligence and Computational Cognitive Science", *in Proc. AISC 2017, 14th Conference of the Italian Association of Cognitive Sciences* (Bologna).

MacKay, D. G. [1992] Constraints on Theories of Inner Speech, in D. Reisberg (ed.), *Auditory Imagery* (Psychology Press), pp. 131–148.

Malsburg von der, C. [1997] The Coherence Definition of Consciousness, in M. Ito, Y. Miyashita and E. T. Rolls (eds.), *Cognition, Computation and Consciousness* (Oxford University Press), pp. 193–204.

Manzotti, R. and Tagliasco, V. [2007] "An Externalist Process-Oriented Framework for Artificial Consciousness", in *Proc. AI and Consciousness: Theoretical Foundations and Current Approaches*. AAAI Fall Symposium, Technical report FS-07-01. AAAI Press, (Menlo Park California) pp. 96–101.

Marchetti, G. [2006] A presentation of Attentional Semantics, *Cognitive Processing* 7(3).

Maslin, K. [2001] *An Introduction to the Philosophy of Mind* (Polity Press), p. 87.

Massimi, M., Ferrarelli, F., Huber, R., Esser, SK., Singh, H. and Tononi, G. [2005] Breakdown of Cortical Effective Connectivity During Sleep, *Science* **309**(5744) 2228–2232.

Mavridis, N. and Roy, D. [2006] "Grounded Situation Models for Robots: Where words and percepts meet", in *Proc. IROS'2006*, pp. 4690–4697.

McCulloch, W. S. and Pitts, W. [1943] "A logical Calculus of Ideas Immanent in Nervous Activity", in *Bulletin of Mathematical Biophysics*, Vo 5, 1943, pp. 115–133.

Miyawaki, Y., Uchida, H., Yamashita, O., Sato, M., Morito, Y., Tanabe, H. C., Sadato, N. and Kamitani, Y., [2008] Visual Image Reconstruction from Human Brain Activity using a Combination of Multiscale Local Image Decoders, *Neuron* **60**(5), 915–929.

Morin, A. and Everett, J. [1990] Inner speech as a mediator of self-awareness, self-consciousness, and self-knowledge: an hypothesis, *New Ideas in Psychology* **8**(3), pp. 337–356.

Musser, G. [2016] Consciousness Creep. *Aeon*, 25 Feb 2016 https://aeon.co/essays/could-machines-have-become-self-aware-without-our-knowing-it.

Nagel, T. [1974] What Is it Like to Be a Bat?, *Philosophical Review* 83, 435-50.

Naselaris, T., Prenger, R. J., Kay, K. N., Oliver, M. and Gallant, J. L. [2009] Bayesian Reconstruction of Natural Images from Human Brain Activity, *Neuron* **63**(6), 902–915.

Newell, A. and Simon, H. [1976] Computer Science as Empirical Inquiry: Symbols and Search. *Communications of the ACM* 19(3) 902–915.

Nii, H. P. [1986] The Blackboard Model of Problem Solving, *AI Magazine* 7(2), 38–53.

Nishimoto, S., Vu, A. T., Naselaris, T., Benjamini, Y., Yu, B. and Gallant, J. L. [2011] Reconstructing Visual Experiences from Brain Activity Evoked by Natural Movies, *Current Biology,* 21(19), 1641–1646.

Pavlov, I. P. [1927/1960] *Conditional Reflexes* (Dover Publications).

Pearson, K. [1911/2007] *The Grammar of Science* (Cosimo Inc.).

Reggia, J., Monner, D. and Sylvester, J. [2014]. The Computational Explanatory Gap. *Journal of Consciousness Studies* 21, 153–178.

Reggia, J, Katz, G. and Huanga D. [2016] What are the computational correlates of consciousness? *Biologically Inspired Cognitive Architectures* Vol. 17, July 2016, 101–113.

Reggia, J. Katz, G. and Davis, G. [2018] Humanoid Cognitive Robots That Learn by Imitating: Implications for Consciousness Studies. *Frontiers in Robotics and AI* 26 Jan 2018 https://doi.org/10.3389/frobt.2018.00001.

Rosenblatt, F. [1958] The Perceptron: a Propabilisitc Model for Information Storage and Organization in the Brain, *Psychological Review* 65(6), 386–408.

Rosenthal, D. M. [2004] "Varieties of Higher-Order Theory" in R. J. Gennaro (ed.) *Higher-Order Theories of Consciousness*, (John Benjamins Publishers), pp. 19–44.

Samsonovich, A. V. [2010] "Toward a Unified Catalog of Implemented Cognitive Architectures", in A. V. Samsonovich, K. R. Johannsdottir, A. Chella and B. Goertzel (eds.), *Biologically Inspired Cognitive Architectures 2010* (IOS Press), pp. 195–244.

Sanz, R., López, I. and Bermejo-Alonso, J. [2007] "A Rationale and Vision for Machine Consciousness", in A. Chella and R. Manzotti (eds.), *Artificial Consciousness* (Imprint Academic), pp. 141–155.

Sanz, R., López, I., Rodríguez, M. and Hernández, C. [2007] Principles for Consciousness in Integrated Cognitive Control, *Neural Networks* 20(9), 938–946.

Schneider, S. and Turner, E. [2017] Is Anyone Home? A Way to Find Out If AI Has Become Self-Aware. https://blogs.scientificamerican.com/observations/is-anyone-home-a-way-to-find-out-if-ai-has-become-self-aware/.

Schneider, S. and Turner, E. [2019] *Artificial You: AI and the Future of Your Mind* (Princeton University Press).

Scott, M. [2013] Corollary Discharge Provides the Sensory Content of Inner Speech. *Psychological Science,* SAGE Journals https://doi.org/10.1177/0956797613478614

Searle, J. R. [1980] Minds, Brains, Programs. *Behavioral and Brain Sciences* 3(3): 417–457.

Shanahan, M. P. and Baars, B. J. [2005] Applying Global Workspace Theory to the Frame Problem, *Cognition* 98(2), 157–176.

Shanahan, M. [2010] *Embodiment and the Inner Life* (Oxford University Press).

Sloman, A. [2010] An Alternative to Working on Machine Consciousness, *IJMC* **2**(1), 1–18.

Sommerhof, G. [2000] *Understanding Consciousness* (Sage Publications).

Steels, L. [2003] Language Re-Entrance and the "Inner Voice", in O. Holland (ed.), *Machine Consciousness* (Imprint Academic), pp. 173–185.

Suddendorf, T., Addis, D. R. and Corballis, M. C. [2009] Mental time travel and the shaping of the human mind, *Phil. Trans. R. Soc. B* 2009 364, 1317–1324. doi: 10.1098/rstb.2008.0301.

Takeno, J., Inaba, K. and Suzuki, T. [2005] "Experiments and examination of mirror image cognition using a small robot", in *Proc. 6th IEEE International Symposium on Computational Intelligence in Robotics and Automation (CIRA 2005)*, pp. 493–498.

Tononi, G., Edelman, G. M. and Sporns, O. [1998] Complexity and coherency: integrating information in the brain, *Trends in Cognitive Sciences* **2**(12), 474–484.

Tononi, G. [2004] An information Integration Theory of Consciousness, *BMC Neuroscience* 2004, 5:42. doi:10.1186/1471-2202-5-42.

Tononi, G. [2008] Consciousness as Integrated Information: A provisional Manifesto, *Biological Bulletin* **215**(3), 216–142.

Turing, A. M. [1950] Computing Machinery and Intelligence, *Mind* LIX no 2236 433–460.

Wand, M. and Schultz, T. [2009] "Towards Speaker-Adaptive Speech Recognition based on Surface Electromyography", in *Proc. International Conference on Bio-inspired Systems and Signal Processing (Biosignals 2009)* (Porto, Portugal).

Zwaan, R. A. and Radvansky, G. A. [1998] Situation Models in Language Comprehension and Memory, *Psychological Bulletin* **123**(2), 162–185.

Index

245

natural language; 12, 65–69, 91, 153, 155, 160, 185, 204, 218
Naselaris, T.; 166
neural correlates of consciousness; 27
neural networks; 46, 81–110, 154, 203
neuron; 22, 24, 81, 27, 41, 81– 86, 91–102, 114, 115, 214
neuron group; 103–110
neuron models; 82, 89
Newcombe, R.; 68
Newell, A.; 2, 3
Nii, H. P.; 177
Nishimoto, S.; 166
nociception; 44

own will; 77

pain; 150, 151, 152, 158, 160, 191, 204, 213, 214, 215, 223, 224, 230
panpsychism; 24, 34
paresthesia; 44
pattern recognition; 84, 86, 176
Pavlov, I.; 58
Pavlovian conditioning; 58, 59, 90, 91, 232
P-consciousness; 29, 30
perception process; 35, 41, 112, 113, 115, 122, 156, 157, 192, 229
perception/response feedback loop; 113–140, 146, 147, 148, 151, 152, 158, 159, 164, 167–172, 187, 206, 208–212, 224, 229, 230
perceptron; 83, 84, 85, 88, 90, 91, 96
phenomenal awareness; 197
phonemes; 108, 110, 127, 128, 171, 172
Physical Symbol System Hypothesis; 2, 3, 5
PID controller; 145
pitch; 127, 207
Pitts, W.; 83
pixel map; 118, 119, 132

Plato; 21
pleasure; 27, 39, 71, 74, 75, 76, 77, 119, 150, 151, 152, 160, 187, 213, 214
prediction; 110, 112, 114, 119, 120, 121, 125, 129, 148, 153, 157
priming; 115, 116, 117, 157, 212, 227, 228, 229
property dualism; 24, 25, 26
proprioception; 44

qualia; 13–20, 28–41, 45, 47, 55, 72–75, 122, 156, 163, 185, 186, 192–197, 201, 229, 230

Radvansky, G. A.; 68
reality; 62
reasoning; 31, 45, 62, 153
recognition; 48, 49, 50, 84, 86, 125, 127, 172, 204, 207, 209
Reggia, J.; 27, 45
reportability; 29, 39, 161, 162, 186, 187
Rosenblatt, F.; 83
Rosenthal, D. M.; 70
rouge test; 199
Roy, D.; 68
rubber hand experiment; 54
Rumelhart, D. E.; 48

Samsonovich, A. V.; 155
Sanchis, A.; 173
Schneider, S.; 197
Scherzer, P.; 31, 32, 33
Scott, M.; 41
Searle, J.; 5, 6
self-awareness; 69
self-concept; 199, 214, 215, 223
self-consciousness; 34, 43, 44, 69, 80, 154, 198–200, 214
self-explanatory information; 8, 12, 13, 20, 31, 35, 45, 46, 154, 190, 198, 230